California Natural History Guides: 17

THE CLIMATE

OF

SOUTHERN CALIFORNIA

BY

HARRY P. BAILEY

UNIVERSITY OF CALIFORNIA PRESS

BERKELEY AND LOS ANGELES 1966

UNIVERSITY OF CALIFORNIA PRESS
BERKELEY AND LOS ANGELES, CALIFORNIA

© 1966 BY THE REGENTS OF THE UNIVERSITY OF CALIFORNIA
LIBRARY OF CONGRESS CATALOG CARD NUMBER: 65-26070
PRINTED IN THE UNITED STATES OF AMERICA

CONTENTS

[3]

INTRODUCTION

The subject of the climate of southern California is one of more than usual interest, if we are to follow ordinarily reliable indications of popularity. Comedian Bob Hope has wisecracked for years about California weather events that run contrary to its stereotype as a kind of extra-tropical paradise. The late Monty Woolley, carrying out the same tradition, once remarked that only in southern California could one freeze to death under a rose bush in bloom. Perhaps this kind of nonsense is a necessary antidote to the saccharine, almost worshipful accounts that are to be seen in the writings of early travelers, some of whom were imbued with the notion that what must be good for trees (as evidenced by the size and age of the Giant Sequoia) must also be good for man.

In a more serious vein, however, we note that the routine publications of the U. S. Weather Bureau dealing with California climate usually run out of print promptly, and that the California issue of the *Climates of the States* series is no longer to be had, which is one of the reasons for using space in this little book to present tabular data in the appendix. One climatologist, who is the author of a well-known volume on the climates of entire continents, has become so impressed by the popular interest in the climate of California that in the later editions of his book he has interpolated a separate chapter on California alone, the only part of Anglo-America to be treated separately from its continent.

A decade ago, in a penetrating analysis of the migration of people to California (which has now attained the rank of one of the great migrations of all time), Edward Ullman, a geographer of note, concluded that for the first time in human history large numbers of people could change their location to gain amenities, rather than for economic advantage. Of the amenities that California has offered, Ullman noted that none is more important than the attraction of its climate. And since statistics of comparative regional increase show that southern California has mushroomed most spectacularly, its climate is a legitimate focus of interest.

What is the essence of the lure of the climate of southern California? I've asked that question of many people. My grandfather gave his answer in terms of the justifiably high expectation that rain would not spoil an outing scheduled well in advance—a reaction natural enough for a retired preacher who had seen many a church social spoiled in the rainier climate of eastern Pennsylvania. A student who was also a mother of active, runabout boys gave quite a different aspect to the same point: to her the chief climatic advantage of Los Angeles over New York was relief from the dread that lighting would strike down her children while at play.

Other kinds of answers have been given to the same question. The mild climate of coastal California has seemed to many a kind of deliverance from the tyranny of the elements, not only because of the absence of snow and rarity of rain, but also because of relief from extremes of cold and heat. Still another group places emphasis not on the temperateness of the climate but upon its variety, with considerable change obtainable by relatively short travel from one location to another. Inevitably, the fact of variety leads to partisan regional preferences, a discussion that need not concern us here other than to establish that variety does indeed exist in southern California, and to a degree unknown on the more level terrains east of the Rocky Mountains.

GEOGRAPHIC SETTING

The momentary state of the atmosphere we call weather; its long-term state is climate. Since weather is variable, climate embraces a multitude of unlike events, and no simple method has ever been found to express all of them with complete accuracy. The climatologist knows beforehand, then, that in the eyes of some he will commit sins of commission or omission; he must find his satisfaction in bringing any description to a partial state of order.

Southern California, we note, is at the margin of a continent across which large transfers of air take place at some times, and little at others. Thus the "marine effect" on climate is strongly variable. The climate usual at the coastline, in other words, moves far inland when sea air in considerable depth moves strongly on land. It is also possible for land air to sweep out to sea. When that happens even the coastline may experience hot, dry air more typical of the desert interior.

The difference in the state of the air over water and land is considerable, especially in summer when land surfaces are much warmer than the sea surface. The separation between the two is more than a matter of distance, for the terrain itself tends to divide the air of southern California into either the sea or the land variety. A glance at figures 1 and 2 will confirm this fact. If we accept the 35th parallel as the northern boundary to the region, southern California contains two prominent mountain systems. The first, stretching inland from Point Arguello for more than 200 miles, is the Transverse Range. The second, the Peninsular

Range, extends northward from the Mexican border. Both ranges gain altitude inland; their highest peaks are only 20 miles apart and are located close to the intersection of the axes of the two ranges, near the town of Beaumont.

Thus, a triangular coastal sector exists seaward of the Transverse and Peninsular Ranges, of which the hypotenuse is the coastline. For our purposes, this sector will be called the Lowland of Southern California. Open to the sea for more than 200 miles, this Lowland is readily invaded by marine air which not only maintains fairly temperate thermal conditions throughout the year, but also brings with it moisture, which if precipitable at all is likely to leave most of its rainfall on coastal slopes.

The Lowland of Southern California contains less than one-half the area of the state south of latitude 35, but it is the home of most of southern California's people, and the site of all of its large cities. For the Indian population too this coastal sector was much more hospitable than the interior, and their villages were thickly clustered near water, along the coast and in certain valley locations. Small wonder that in the past the Lowland of Southern California was synonymous with the notion of the geographic entity of southern California itself.

Beyond the mountain perimeter of the Lowland lie broad expanses of desert, made desertic principally by their inaccessibility to moist air off the sea. The same exclusion allows summer warmth to reach a high level, which in turn contributes to the aridity by increasing rates of evaporation. The abundant sunshine of the region early attracted the attention of farmers who readily found that their crops would grow both summer and winter, provided water were available. More recently, the region has gained in recreational appeal to the urban population of the coast, all the more so as air pollution in the large cities has occasionally reached objectionable levels. The same mountain barriers which intensify the difference between coast and interior in terms of cloud and moisture also intensify the difference in quality of air. Taking advantage of the lack of clouds, large facilities for testing of aircraft and other equipment have been located in the California deserts in postwar years. No longer it is possible to think of southern California exclusively in terms of its coastal sections.

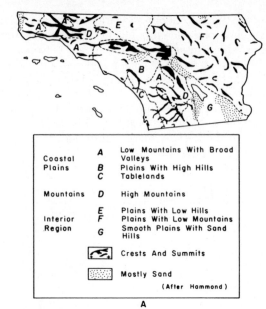

Coastal Plains	A	Low Mountains With Broad Valleys
	B	Plains With High Hills
	C	Tablelands
Mountains	D	High Mountains
Interior Region	E	Plains With Low Hills
	F	Plains With Low Mountains
	G	Smooth Plains With Sand Hills

Crests And Summits

Mostly Sand

(After Hammond)

A

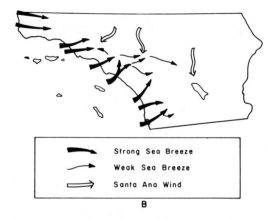

Strong Sea Breeze

Weak Sea Breeze

Santa Ana Wind

B

Fig. 1. A) Land-surface forms of southern California (after Hammond); B) Major pathways followed by sea and land air.

LAND-SURFACE FORMS OF SOUTHERN CALIFORNIA AND AIR FLOW

At first glance a terrain map or relief model of southern California seems to be a bewildering mass of detail, with little pattern or repetition. For climatological purposes, however, a good deal of simplification is permissible, as in the small-scale map of land forms (fig. 1A). Appropriate place names appear on the more detailed location map, figure 2.

Few large areas in southern California are free of mountains, and from the air, on a clear day, mountains are always in sight. The ridge crests of the most prominent are shown as elongate, black lines in figure 1. Although the Transverse and Peninsular Ranges create an east-west and north-south highland within southern California, those highlands are in themselves complex, composed of many individual ridges departing in orientation from their collective wholes. In the Peninsular Range, particularly, the mountains show disparate trends; in the south many ridges descend to the sea at right angles to the main north-south axis. Others are arranged in rows, with a northwest-southeast trend, markedly so in the Santa Ana Mountains. Together with other hills and mountains aligned on the same axis, they stand out well in front of the main body of the Peninsular Range: so far, in fact, that they create the customary dividing line between the upper and the lower basins of the Santa Ana and San Gabriel Rivers, two of the three major river systems in the northern part of the Lowland of Southern California.

[11]

Fig. 2. Location.

The third, the Los Angeles River system, is likewise divided into lower and upper basins by the east-west trending Santa Monica Mountains.

The Lowland of Southern California is thus not uniformly open to the influx of sea air. Areas close to the sea naturally receive maximum influence from air that has shortly before been over water; as such air moves inland it tends to follow uphill those same avenues that water follows downhill, with the difference that air easily surmounts minor obstacles. In figure 1B the major channels for influx of the summer sea breeze are shown by broad, solid arrows. Where the arrows are thin, the underlying terrain reduces but does not eliminate the sea breeze completely.

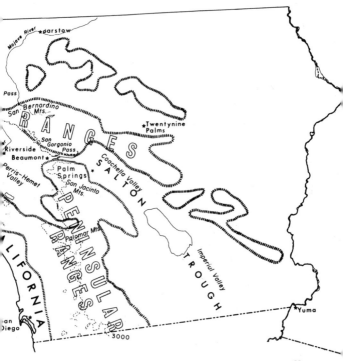

From figure 1B it is clear that, although river valleys opening to the sea always allow air to move inland, deep penetration occurs only where the rise of extensive plains surfaces is gradual. Distance from the sea is only an approximation of the actual distance traveled by the sea breeze, moving as it must over relatively low ground. For example, the western San Fernando Valley, one-third as far from the ocean as the upper Santa Ana Basin, receives only a little more sea air. The San Fernando Valley is open to such influence circuitously through just one narrow gap, the Glendale Narrows, while the Santa Ana Basin receives sea air over two relatively direct and broad avenues: the canyon of the Santa Ana River, cut through mountains of the same name, and the undulating piedmont slopes flanking the Transverse Range. In contrast, the Perris-Hemet plain, fully as extensive as the San Fernando Valley, and adjacent to the upper Santa Ana Basin, is not freely open to the summer breeze by its altitude (averaging about 1500 feet) and surrounding highland barriers.

Valleys of small dimensions flanked by highlands receive sea air in such small volume that it rapidly mixes with interior air and so loses identity, even close to the sea. Such is the case with the Santa Clara River, which creates a broad alluvial apron under full marine influence near the ocean, but narrowing of its valley no more than 15 miles upstream to a width of less than a mile prevents the upper part of its basin from receiving important amounts of sea air under usual sea breeze conditions. So it is with the many small valleys that rise from the sea between San Diego and Oceanside.

Under normal conditions, the inner limit of the sea breeze is safely set at the mountain perimeter of the Lowland of Southern California. Usually no more than 3000 feet thick at the outset, and undergoing constant attrition by mixing as it moves inland, the sea breeze necessarily reaches its "coastline" close to the 3000-foot contour, which has been plotted on figure 2, along the front of the Transverse and Peninsular Ranges.

Beyond those ranges is situated a large region to the north and east that is hot in summer, and dry all year around. Even the stormiest and most buoyant winter-type storms are so disturbed in their passage over the mountains that rain-making (and snow-forming) processes are greatly affected. This disturbance reduces the wetness of the lower slopes of the mountains facing north and east and of large areas beyond. Such areas are aptly called "rain-shadow desert," and are found from Mexico to Canada in the lee of the Pacific mountain systems.

The desert north of the Transverse Range is known as the Mojave. In its western portion it is relatively open and flat; basin floors there average 2500 to 3000 feet in altitude. To the east of the Mojave River, the surface rises, and is interrupted by many mountains. Here, and farther west, the deserts are called "high" by the Weather Bureau, distinguishing their cooler and slightly wetter conditions from the eastern Mojave and the Salton trough, called "low deserts" for closeness to sea level.

The Salton trough is the lowest, flattest, and hottest of the desert areas of southern California (not including Death Valley). To its west the Peninsular Range rises impressively. On the east and north less prominent hills and mountains, extensions of the Transverse Range, nearly complete its highland rim. Although it is open to the south, which means this valley occasionally receives warm, humid air moving northward from the Gulf of California, even this advantage is insufficient to relieve the prevailing aridity of its transmountain location.

It is a curious fact that the highest and lowest areas of southern California are in relatively close proximity. As mentioned, the Transverse and Peninsular Ranges rise inland; their highest peaks (San Gorgonio Peak, 11,485 ft.; San Jacinto Peak, 10,805 ft.), are separated by the floor of San Gorgonio Pass, more than a mile below them on the vertical. This pass leads southeast to the Salton Sink, whose floor is 275 feet below sea level. In the opposite direction, the trend of San Gorgonio Pass leads to Cajon Pass, the lowest and widest break in the Transverse Range. Thus, within the space of forty miles, two major passes link three important plains areas of southern California; the western Mojave, the Coachella (the local name applied to the upper end of the Salton trough), and the upper Santa Ana basin.

The same passes and lowlands are interconnected

atmospherically. On a warm summer day, the sea breeze reaches the upper Santa Ana basin, as noted previously, and then moves in some quantity through Cajon Pass and San Gorgonio Pass, where it is overwhelmed by the immense pools of hot air over the Mojave and Coachella.

However, reverse movements of great magnitude occasionally take place in the cooler months of the year, when cold air piles up in the continental interior, and laps against the desert face of the Transverse Range. Then the mountain passes are inundated by great streams of air moving swiftly from the interior toward the Lowland of Southern California. Cajon Pass in particular is an avenue for cold air making its way downhill from the elevated western Mojave. Under the most active conditions, the entire crest of the Transverse Range is crossed by strong northerly and northeasterly winds, creating a windstorm over all of southern California.

It is apparent that the climate of southern California is greatly affected by the ease with which sea air and continental air can move to any given locality, which largely causes the regional differences in temperature, precipitation, cloudiness, fogginess, sunshine, and windiness. In general, the climate becomes warmer, drier, and more sunny as distance from the coast increases. These tendencies, though, are true only for lowlands. If the sea-to-interior movement involves crossing mountains, as it must with only a few exceptions, then the effects of altitude are also encountered.

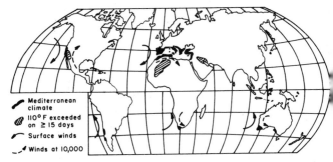

Fig. 3. World distribution of the "Mediterranean" climate and areas of extremely high summer temperatures.

THE CALIFORNIAN CLIMATE ABROAD

Despite the fact of internal variety, it is fair to say that the climate of southern California is sunny, warm, and rather dry, with the meager rain that does occur falling during the cooler months of the year. These characteristics define what is often termed the "summer-dry subtropical" climate, even better known as the "Mediterranean" climate because of the great length of the Mediterranean coastline which also displays similar conditions.

Figure 3 shows that several regions of Mediterranean climate also occur in the Southern Hemisphere. Now, most parts of the world that are dry in summer are even drier in winter, and thus are clearly desert or semi-desert. In the case of the Mediterranan climate, wet winters intervene between the dry summers, a combination so rare that it is encountered on less than 3 percent of the earth's land surface. It is highly significant that all areas of Mediterranean climate are located between the 30th and 45th parallels of latitude, and are on the western borders of the land masses of which they are a part. No more remarkable example exists of system and symmetry in climatic distribution.

The explanation for such a scattered but patterned distribution lies in the largest elements of the planetary air and water circulation. California, and all its counterparts abroad, are dominated in summer by large, placid pools of air formed at sea, indicated on weather maps as large, elliptical swirls of air classed together as the system of subtropical anticyclones. They are shown schematically in figure 3 by the solid arrows. These anticyclones, perhaps 2,000 miles across at the widest point, are tangent to and to some extent overlap the west coasts of all continents somewhere in the latitude range of 30° to 40°, and lend to the land in their vicinity nearly absolute drought in summer months.

In winter, in contrast, these same cells of storm-free air shrink in size, and variably retreat equatorward and seaward, which allows storms originating in higher latitudes to sweep across the same coasts that are so calm in summer. The contrast between the rainy winters and dry summers of southern California would not be complete without mentioning one further aspect of anticyclonic influence: the air in anticyclones is in slow descent, and as a consequence warms from the effects of compression as the upper parts sink into denser, lower layers of the atmosphere. Warming stops short of the surface, for the air above the sea has a temperature closely similar to that of the water, and the tumbling action of the air above the waves carries to perhaps the 2000-foot level.

There, at that distance above the sea, the cool, humid marine air meets the warm descending air, and creates the temperature "inversion" for which southern

California has become so well known. That is, although the temperature decreases with altitude at a normal rate for the first 2000 feet or so, it increases fairly abruptly as you move above the marine air into that overlying, as shown in figure 4. The zone of contact of the two unlike air masses is often clearly visible to the eye, the sea air usually being hazy, the air above clear. Furthermore, since the bottom air is cooler than that immediately overlying, it gains in density as well. This gives so little opportunity for vertical movement above the level established at the base of the temperature inversion that the chance for summer rainfall is virtually eliminated.

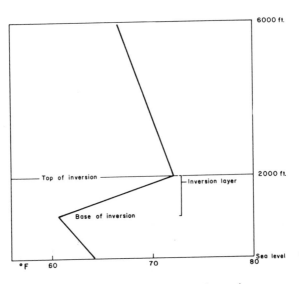

Fig. 4. A typical temperature inversion over the southern California coast.

If it is the layered structure of air over the areas of Mediterranean climate which must be blamed for their lack of summer rain—usually accounted as a type of climatic demerit—we note that such a lack has distinct advantage in terms of human comfort. Most of the land area of the world experiences a parallel trend in regimes of rainfall and temperature, with the result that the peak of summer heat is matched with the peak of humidity. But where the warmer half of the year is attended by little or no rain and relatively dry air, the discomfort of summer heat to human beings, or to any animal with sweat glands, is reduced.

Even if the humidity factor is disregarded, the coastal parts of the summer-dry subtropical regions are free from extremes of temperature in a degree unusual for localities close to sea level. For example, on the basis of a scale proposed in a recent study, where 100 stands for complete freedom from extremes of temperature, coastal California rates from 60 to 70, whereas all of the American midwest and northeast drop well below 50, as does southern Florida.

With distance inland, the temperate character of the Mediterranean climate is gradually reduced; in fact, as seen in figure 3, adjacent to them are the hottest areas of the world, as measured by the frequency of extremely high temperatures. This circumstance is not unrelated to the preceding discussion. The presence of clear, dry air in the upper parts of the subtropical anticyclones allows solar heat to penetrate with great intensity toward the ground. Although some of that heat is lost by reflection from low-lying clouds, and some is utilized in the evaporation of moisture—processes that effectively reduce air temperatures over the ocean—those losses are not as large inland, with the result that great quantities of heat are poured on the ground, day after day in the summer. This would not be possible if astronomical factors were not also favorable toward the development of high summer temperatures. Latitudes between 30° and 40° receive more heat from the sun in summer months than does the equator, a circumstance which, in combination with nearly cloud-free skies of the lower Colorado River Valley, North Africa, and the Persian Gulf, goes a long way in the explanation of their great summer heat.

The relation between summer afternoon temperatures (actually, the mean maximum temperatures in July) in southern California and distance from the sea is shown in figure 5. For the sake of objectivity, the distance from the sea represents the shortest line connecting a given station with the coastline, even though we know movements of the sea breeze depart considerably from a straight-line path. Since the trend line shown in figure 5 approximates a line of best fit, we can hypothesize that stations warmer than predicted by the trend line are places with less than average access to sea air, while the opposite should be true of stations cooler than the trend line. In that light, Ojai, Elsinore, and Palm Springs are interpreted to have poorer than average ventilation from the sea

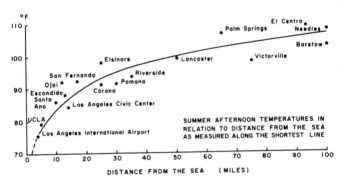

Fig. 5. Increase of temperature with distance inland on an average summer afternoon.

breeze, while downtown Los Angeles, Corona, and Riverside are treated to more than their share of the sea breeze.

Beyond the Transverse and Peninsular Ranges, distance from the sea is probably less important than altitude as a control of summer temperature. Thus, Lancaster and Victorville, (2300 feet and 2900 feet respectively) are cooler than stations low in the Salton trough. In any event, the graph is terminated at a distance of 100 miles from the sea, and all points further than that are treated as if they were 100 miles from water, on the supposition that the marine influence is negligible so far from the coast. The maximum afternoon temperature predicted from the trend line is 107°, which is exceeded slightly in the means observed in the Imperial Valley, and the Colorado River Valley below Hoover Dam.

CLIMATIC REGIONS
OF SOUTHERN CALIFORNIA

Figure 6 shows seven types of southern California climate, chosen to portray regional differences important to both natural and man-made landscapes. Appendix B supplements the map: table 1 by defining the limits of each climatic type, table 2 by giving data for at least one station in each type, chosen for the completeness of record. The climatic map, and supporting tables, were based largely upon data from records and instruments supervised by the U. S. Weather Bureau from 1931 to 1952, inclusive. Certain data in table 2 are from older records, and not all items are reported for as long as 22 years.

Our examination of figure 6 starts with looking at the pattern as a whole. The separation of the coastal sector from the interior, so clearly marked on the ground by the Transverse and Peninsular Ranges, is expressed climatically by the dotted temperature line. Along the path of this line there is a difference of 30 degrees between the temperature means of the warmest and coldest months, based on an average of day and night conditions. Seaward of that line, seasonal differences of temperature are less than 30 degrees, and toward the interior they increase to 40 degrees or more at the Arizona border. The coastal sector is also far wetter; the large area of desert climate shown in figure 6 has been set aside as a region with less than 10 inches of precipitation annually. No part of the coastal sector is quite that dry; being cooler and

[23]

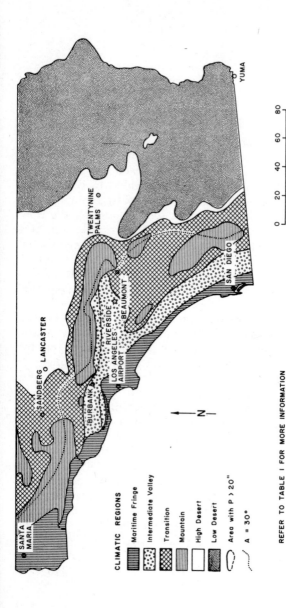

CLIMATIC REGIONS

- Maritime Fringe
- Intermediate Valley
- Transition
- Mountain
- High Desert
- Low Desert
- Area with P > 20"

A = 30°

REFER TO TABLE I FOR MORE INFORMATION

Fig. 6. Climatic regions of southern California.

cloudier as well, coastal slopes lose less water by evaporation. Consequently watersheds of the Pacific slope send streams that reach the sea, if only with intermittent flow. The slopes leading to the interior, in contrast, are unable to form rivers, even in flood, that make the full journey to the sea. The largest of them, the Mojave, has a sizable flow as it leaves its source region, the San Bernardino Mountain section of the Transverse Range, but it evaporates to dryness in its lower course. The history of rainfall at Bagdad, in the eastern Mojave, clearly shows why evaporation gains the upper hand. Bagdad holds the record in the United States for the greatest number of consecutive days without rain: 767.

The coolest and wettest type is the Mountain climate, defined as a region where the coldest month of winter is below 50° and annual precipitation exceeds 20 inches. The Low Desert is the warmest and driest region of southern California, a striking contrast to nearby mountains. Palm Springs, for example, with an altitude of 411 feet and a mean annual precipitation of 7 inches is next to the massive slopes of the San Jacinto Mountains whose summits, two miles high, are wetted by more than five times as much rain, and are often snow-capped in winter.

THE MARITIME FRINGE

It is logical to start our examination of individual climatic types with the Maritime Fringe, which is the first land to be crossed by air coming off the sea, the usual type of circulation. Only in the coastal strip bordering the sea do we find stations with records of summers cool enough and winters warm enough to satisfy the stipulation that the warmest month must be below 72° and the coldest month above 50°. Even afternoon temperatures in July and August are comfortably cool, and no other part of southern California is quite as warm on January mornings. The longest growing season of all of the Southland is to be found in the Maritime Fringe.

On a winter day in southern California, when the sun is above the horizon only 10 hours a day and is low in the sky even at midday, the amount of heat received from the sky from sunup to sundown is only half that received during the 14 sunlit hours of a summer day, when the sun nearly reaches the zenith at noon. This kind of difference, of course, is the reason for the seasons in all midlatitude localities; in them the only way that temperatures can remain relatively constant throughout the year is to buffer heat intake and outgo. The buffering process is carried out primarily by the oceans, which have a large heat storage capacity. This capacity is transmitted most effectively to the atmosphere by charging it heavily with water vapor, which leads, preferably, to a complete cloud cover. The price of thermal equability in midlatitude locations, it must be emphasized, is to forego clear skies and dry air. This price is paid most heavily in oceanic areas and islands such as the Aleutian and Faeroe Island areas.

Coastal southern California is not nearly as cloudy as the areas just mentioned; even the cloudiest month at Los Angeles Airport has less than half its days ranked as "cloudy" (seven-tenths or more of the sky covered by clouds). Heavy fog occurs there on the average of 53 days of the year, mainly in the cooler half of the year. Santa Maria's 88-day frequency of heavy fog is concentrated in late summer, but fog decreases southward, being noted at San Diego's waterside airport on only 30 days a year, mainly in winter. However, a "day" with fog in this case means only that heavy fog (visual range less than one-quarter mile) occurred sometime during the day. Even with the exacting demands of jet-powered aircraft it is unusual for fog to close down an airport for more than a few hours at a time. When such conditions do exist, freeway, harbor, and even street traffic is greatly reduced.

The distribution of overcast skies and fog has never been mapped in detail in southern California. To the resident, their patterns show infinite variation in detail, at the same time that certain features are fairly repetitive. Low ground near the coastline is certainly most affected by clouds and fog that have formed over the sea. In fact, it is not far wrong to say that the Maritime Fringe itself defines fairly well the area most affected. Widest at the north, where ocean temperatures are low, the cool Santa Maria and Lompoc Valleys are comparatively cloudy and foggy. From Point Arguello south to Santa Monica, hills and mountains rise from the sea almost uninterrupted by major valleys, and so the Maritime Fringe extends landward only a few miles from the water's edge. The warmth generated by south-facing slopes often dissipates clouds lying against them, and their upper parts are bathed in such warm air in summer that they usually stand out boldly above whatever clouds cover the beach. Over the Los Angeles lowland sea air flows with relative ease, and so there the Maritime Fringe reaches nearly as far inland as the Civic Center of Los Angeles and Santa Ana, 10 miles or more from water. South of Newport Beach, however, the coast is cliffed, and the terrain takes the form of tablelands through which sea air can move only in the narrow ribbons provided by confined river valleys. Consequently, the Maritime Fringe is again narrow south of Newport Beach.

The sea air crossing the Maritime Fringe is quite moist, as is to be expected. Relative humidities in early morning and late afternoon are above 60 percent, and reach 90 percent at 4 A.M. in summer months at Santa Maria. Early morning hours in summer are also the most humid in Los Angeles, where the relative humidity exceeds 80 percent at 4 A.M. May through

August at the Civic Center, and for a considerably longer period (March through October) at the airport, which is only a few miles from the sea. In the heat of the day, however, the relative humidity drops at least 20 points from early morning levels, and is lower still during heat waves.

In spite of these cloud and humidity conditions, one should not assume that the Maritime Fringe is a region without sunshine. It is, in fact, sunnier than many parts of the United States. From 60 to 70 percent of daylight hours are sunny there, which is consistent with records which show that about half the days of the year are ranked as "clear" (three-tenths or less of the sky covered with clouds). Intermittent sunshine on partly cloudy and cloudy days also adds to the sunshine total.

Also, it would be expected that cloud, fog, and humidity would decrease when moving inland across the Maritime Fringe; sunshine, of course, would increase. Those distributions (stated conditionally because so few data are available, especially for sunshine) are not only confirmed by the data we have, but are also consistent with the notion that sea air moving landward merges gradually with the drier, clearer air of the interior.

There is little doubt that the seaward portion of the Maritime Fringe offers great attraction to many people, especially its favored situation with respect to air pollution. Not only is the air usually fresh off the sea, but temperature, wind and humidity conditions combine to form an environment unexcelled in terms of human comfort. If we define comfort to lie within a range with a lower limit of 40° in still air—which requires a person to add some outer clothing and to exercise lightly to stay warm—and an upper limit of 75° in moist air, or 90° in windy, dry air—heat enough so that a person would want to rest in the shade, with only light clothing—then no other body of temperature data in the United States promises quite so high a degree of freedom from heat and cold. The records at San Diego Airport indicate that a full 90 percent of all hours there are comfortable, including night and winter. Even animals seem to thrive on it—at least, the death rates of animals in the San Diego Zoo are the lowest in the nation!

Here, as in all non-mountainous parts of southern California, the water supply is inadequate, leaving no alternative except bringing water from other, better-supplied source areas. Rainfall in the Maritime Fringe averages about 15 inches annually, but in some years it is as low as 5 inches, and in very wet years as high as 30 inches or a little more. The geographical distribution of rainfall shows how the terrain influences local averages, with most of the rain on or adjacent to rough uplands, and the least on low, level surfaces. The contrast is well shown in the comparison of UCLA with the Los Angeles Airport, the former receiving 18.4 inches on the average, the latter 12.5 inches. As the two places are only 5 miles apart, it can be deduced that rainfall increases at the rate of about an inch per mile over the gentle slopes separating them. Immediately north of UCLA, however, the Santa Monica Mountains provide rough, hilly terrain 1000 feet and more above the altitude of the campus; their east-west axis appears on figure 6 in the 20-inch rainfall line. The Santa Monicas are not high enough to materially lower winter temperatures, however, and so their influence is felt primarily by increase of rainfall.

INTERMEDIATE VALLEY

The Intermediate Valley type of climate, distinctive in its combination of warm summers with mild winters, occurs only in parts of the Lowland of Southern California beyond the Maritime Fringe. It is a warmer climate than that found in the valleys north of Point

Arguello, where in summer temperatures are depressed a few degrees by the cool water offshore, and in winter by the northerly location. In comparison to the Low Desert the Intermediate Valley type is cooler in summer, and has far more rainfall in winter. Many mountain streams enter these valleys, and were used fairly cheaply for irrigation in the last century by the early agricultural colonies that first established extensive citrus plantings.

The Intermediate Valley climate, with its environment, has supplied the image of southern California most commonly held by residents elsewhere in the United States, and abroad. Blue skies over snow-capped mountains in the distance, sunny orange groves in the foreground, a smiling maiden the center of interest—who has not seen this collage a dozen different ways in advertisements of southern California? San Fernando, Pasadena, Pomona, Riverside, Redlands; all fitted the pattern as late as the 1930's, a pattern now altered, of course, by urban growth.

It is to be expected that the warmest parts of the Intermediate Valley region will be found in its inner recesses. At San Bernardino, 50 miles from the sea, the July mean is in the high 70s, and maxima there in the same month average 20° higher still. Redlands, just as far inland, is a little cooler by reason of its greater altitude (1352 feet vs. 1094 feet) a cooling effect which also lowers the January mean of Redlands to 50.1°, which places it just inside the Intermediate Valley limit. Where winter means descend below 50°, the Transitional climatic type takes over, and winter cold precludes widespread subtropical horticulture. The 50° January isotherm, the mutual boundary of the two climatic types, lies on hill slopes from 1000 to 2500 feet in altitude, and is not found on level ground much above 1000 feet.

The geometry of valleys is a sensitive control over the decline of temperature on quiet, clear nights. On such nights, the surface of the ground cools more rapidly than does the air immediately above it from which heat is removed by conduction, thus creating a skin of cold, dense air next to the ground. By sunrise in the winter, after a 14-hour night, a considerable contrast has developed between the conditions on broad valley flats, as compared with adjacent hill slopes. The valley floor is the colder of the two sites because the cold air over it has no place to go. The cold air along valley slopes has drained away nearly as fast as it was formed, making way for warmer air from the surroundings. The "thermal belts" that flank valley sides have been recognized from ancient times; in California they make the difference between sites with light and heavy mid-winter frosts. To a degree man can aid Nature in mixing warmer upper layers with cold bottom air by the mechanical mixing enforced by wind machines, but in the worst cases only outdoor heating can keep temperatures above damaging levels. Using these methods, grapefruit, lemons, oranges, and even avocados are successfully grown in the Intermediate Valley climatic region, although thermometers at instrument shelter height (about 5 feet) have recorded minima in the low 20s, or even colder.

Precipitation records from Intermediate Valley stations show many mean annual totals in the 20-inch plus category. All such wet stations are located at mountain foot sites, where they receive upgliding motions of air overhead, a vertical component of motion essential to production of heavy rain that sometimes takes place well in advance of a mountain front. The same stations do not show much greater frequency of rain. Rain storms in southern California, in other words, are mainly general in character rather than local. A wet locality is essentially one where it rains harder, not more often.

At the Civic Center of Los Angeles rain falls in

measurable amounts for at least a few minutes on an average of 5 days in December and 6 in January. Pasadena, 9 miles away and closer to mountains, should be even rainier than Los Angeles, and in terms of mean annual rainfall it definitely is: 21.9 inches vs. 14.6 inches. In spite of this, the famous Pasadena Tournament of Roses Parade, held each New Year's Day, has had an amazingly dry history. In the 75 years that records have been kept, rain has only fallen during the parade five times: one time in fifteen. Few impresarios of outdoor events ask that odds against rain be better than fourteen out of fifteen!

Many newcomers to the Intermediate Valley region are more impressed by their problems with heat, rather than with cold or wet. To the commuter faced each day with travelling back and forth in a car that feels like an overheated greenhouse, living in a new house not yet shaded by trees, heat is a significant issue. Air conditioning is becoming a standard built-in feature of new home construction away from the Maritime Fringe. Nearly half the days of July, August, and September reach or exceed 90° at Burbank, over 100° at Redlands. These figures, representative of the cool side and the warm side of the Intermediate Valley zone, clearly show the need for refrigeration of air if indoor comfort is to be maintained throughout the day.

Even some days of winter are warm in southern California. Temperatures in the 80s have occurred in January, the coldest month, and other winter months have had still warmer spells. The Rose Parade of 1918 was held on a day when the temperature reached 86 in Pasadena; the record there for March is 98. Hot spells in winter sometimes occur after rain spells, but freezes follow rain storms at other times. In either case, clear dry air comes from the interior to bring southern California its most brilliant sunshine. The coast is the last to lose its cloudy skies when sunshine arrives, and the first to regain them, and so Inter-

Stratus from the top (at about 2000 ft.), near the coastline.

Haze and stratus, often seen near the sea.

Cumulus formed in air roughened by passing over coastal hills.

Stratocumulus (stratus in the process of forming or dissipating).

PLATE 1. COMMON CLOUDS IN THE MARITIME FRINGE.

Cirrostratus, with halo and parhelion (mock sun) faintly visible.

Cirrus in the forefront of the storm.

Altostratus indicating gathering storm.

Altostratus and altocumulus, typical of fading storm.

PLATE 2.
CLOUD SEQUENCE ATTENDING PASSAGE OF WINTER STORM

Lowering altostratus, rain imminent on coastal side of mountains.

Storm clouds forming on desert side of mountains, clearing on coastal side.

After the storm in the mountains.

Post-storm Santa Ana winds in the desert raising dust (near Palmdale).

PLATE 3.

CLOUD SEQUENCE ATTENDING PASSAGE OF WINTER STORMS.

Lenticular altocumulus formed by uplift of northerly flow over the San Gabriel Mountains.

Altostratus formed in moist northwesterly flow, evaporating in lee of Transverse Range, in April.

Post-storm cumulus formed over mountains in still-moist air.

Sunset in the San Fernando Valley, colors on post-storm altocumulus.

PLATE 4.
CLOUD SEQUENCE ATTENDING PASSAGE OF WINTER STORMS.

Sun at the beach, cirrus to the north.

Tangled cumulus formed over the Santa Monica Mountains in westerly flow.

Disturbance at sea.

PLATE 5. THREE MOODS OF WINTER.

Air from the northwest forming clouds against the Transverse Range in June (compare with April).

Cumulonimbus (thunderclouds) forming in summer over high mountains in distance.

August sky during Sonoran weather type.

August rain in west Los Angeles.

PLATE 6. CLOUDS OF SUMMER.

Los Angeles basin, light smog, looking west from Black Mountain (near San Jacinto Peak).

Heavy smog, same view; the smog has ascended mountain slopes to the 8000-ft. level.

Looking over moderate smog toward the San Gabriel Mountains.

Stratus extending inland to San Gorgonio Pass over the Los Angeles basin.

PLATE 7. STRATUS AND SMOG.

Stratus evaporating in San Gorgonio Pass.

Cross-section of the marine layer, topped by stratus at the 8000-ft. level (April), 40 mi. inland.

Smoke from the Bel-Air fire (Nov. 6, 1961).

West wind funneling through San Gorgonio Pass.

PLATE 8. STRATUS AND SMOKE.

mediate Valley climate is reckoned as sunnier than the Maritime Fringe, as well as being warmer in summer. It is a curious fact, however, that the instrument registering sunshine at Riverside has never recorded intensities as great as those occasionally observed at Los Angeles Airport. Presumably the atmosphere close to the sea is purer than it ever becomes inland, even though cloudiness close to the sea reduces the average of all readings.

TRANSITION AND MOUNTAIN CLIMATES

Transition and Mountain climatic types occur on the highland backbone of southern California. The highest mountains have more than 20 inches of precipitation annually, on the average, and it is the areas enclosed by the 20-inch rainfall lines that stand out on the climatic map (fig. 6) as islands of the Mountain type. Surrounding them are drier, and generally lower, areas that are called Transitional. Everywhere they are cool in winter; the 50° isotherm for the mean temperature of January separates the Transition type from the Intermediate Valley type, where winter means are warmer than 50.

Running through the areas of Mountain and Transition climatic types is the temperature line denoting 30° of difference between the means of the warmest and coldest months of the year. The 30°-difference isotherm follows closely the crest of the Peninsular and Transverse Ranges, and so is a climatic approximation of the terrain-marked dividing line between coast and interior. In consequence, the Transitional climatic type, which straddles the isotherm, has been divided in table 1 into parts A and B, standing for coastal and interior phases, respectively (sparse data prevent a corresponding division of the Mountain climatic type).

Sharper seasonal differences in temperature, drier air, and few clouds give the interior phase of the Transition climatic region a definitely semiarid aspect, poorly sampled in the data assembled for this study because so few records are available. It lies in a broad ribbon along the interior faces of the Transverse and Peninsular Ranges, midway between valley floor and ridge crests, where the terrain is rough and steeply sloping, for the most part. Only if the summit areas are quite dry and extensive, as on Bald Mountain in the Tehachapi Mountains—the site of Sandberg, the Transition station with the most complete data—or if elevated valley floors receive between 10 and 20 inches of precipitation, as at Beaumont, do we find people living in the Transition zone.

Sandberg, despite its altitude of 4517 feet, receives only 12.5 inches of precipitation annually, as a result of a shaded position in the lee of numerous ridges intervening between it and the sea. It is cool in winter, when snow is to be expected, but a reliable snow cover fails to develop. The daily range of temperature is strikingly low in winter, as at Mount Wilson, partly because of the winter wind. There is little chance for sun-basked slopes to communicate their heat to overlying air, as they do in summer, for the restless winter air sweeps quickly by California's mountains, bringing them much of the influence of the free atmosphere.

Snow is a usual winter event at altitudes over 4000 feet, and patches of snow last through spring into early summer on high, shady north slopes. Both observation and theory lead to the conclusion that the rain received on the lowlands of southern California originated in clouds far above as flakes of snow. The winter "rains" could then be properly called "snows." As the snow falls, it turns into rain, melted in the lower layers of the atmosphere where temperatures are above the melting point. In the cooler half of the year this takes place at some level below 8000 feet, usually as low as 4000 feet in midwinter.

Two major highways of southern California ascend above the 4000-foot level in crossing the Transverse

Range, Interstate 5 at Tejon Pass (4183 feet) near Sandberg, and Interstate 15 at Cajon Pass (4301 feet) 10 miles northeast of San Bernardino. Several times each winter motorists are inconvenienced by the necessity to use tire chains until a safe surface is re-established on those very busy thoroughfares.

There is usually snow cover suitable for winter sports, at least briefly, at altitudes above 6000 feet in the higher mountains. I hiked to the summit of Mount Wilson in the memorable January of 1933, when five feet of snow stood on the level on the Observatory grounds, and kept the then-unpaved road to Pasadena closed for weeks. In other years, however, I've found it possible to take the modern road to Mount Wilson in midwinter, even on a vehicle as sensitive to surface as a motorcycle.

The average annual snowfall at Mount Wilson is 41 inches, distributed in small amounts over half a dozen storms or so, with the result that the ground often clears between storms. A snow cover is subject to variation not only because of snowfall, but of temperature as well, and few parts of the mountains of southern California can guarantee a snow season. The chancy, uncertain nature of the snow cover has led to great pressure for the modification of the San Gorgonio Wild Area to permit commercial ski facilities on its north face, where at altitudes of 10,000 feet and above snow often persists when it has largely disappeared from lower, warmer sites. Thus in the search for snow fields (a scarce commodity locally) man has discovered a slope upon which glaciation developed in the Ice Age, perhaps the only such slope in southern California.

Sandberg and Mount Wilson are high enough to be unaffected by low-lying clouds; hence, except for the stormy periods (less than 50 days of the year), they are bathed by clear and unusually cloud-free air. The Director of the Mount Wilson and Palomar Observatories informs us that astronomical observations are carried out on those mountaintop sites on about 300 nights a year, and that solar photography on Mount Wilson has been possibe 305 to 325 days a year. This favorable record was anticipated more than half a century ago by George Ellery Hale, the founder of the Mount Wilson Observatory, and was a leading reason for his choice of southern California as a site for its large instruments.

Humidity is prevailingly low in the air surrounding the mountains, and descends to exceptionally low levels during the windy episodes bringing air from the north, the "Santa Ana" winds well known in southern California. At its driest, such air carries so little water vapor that standard instruments and tables barely detect its presence. An observer on duty at Sandberg in the 1930's relates the problem that arose during a Santa Ana episode, when his instruments indicated a humidity so low that he reported a dew point far in the sub-zero range (the dew point is the temperature of saturation—the temperature of a chilled flask, for example, that would barely cause a film of moisture to form on the exterior of the flask). It so happened that this particular observer was being monitored that day in the checking procedure employed to reduce errors of observation and transmission of weather data. The checker, seeing that the dew point was far lower than any in his experience, sent word back that the dew point was in error, and for it to be recomputed. This the observer did, with little change in the result. Again he was told that the dew point was erroneous. At this point the hapless observer realized that he would have to do something if he were to pass the checker, and so his third transmission gave figures more agreeable to the eye of the inspector (if less accurate)—still low enough to convey the information that the air was very very dry indeed!

The Transverse Range to the east of Mount Wilson and the Peninsular Range are more exposed to continental influences, and are somewhat more extreme with regard to temperature. Big Bear Lake, at an altitude of 6800 feet, and surrounded by even higher ground, has recorded –25°, and several other mountain stations in the general vicinity have recorded zero or lower.

In contrast, summer temperatures are comfortably warm everywhere in the mountains. A temperature of 100° or more has been recorded at mountain stations up to nearly the 6000-foot level, but monthly means do not exceed 75° and are generally lower in the 5000-6000-foot range. Several times each summer, more often in August than in July, spells of cloudy weather appear in which thunderstorms develop. The rain they bring is scanty, and lightning is a potent source of forest fires, for the woody material of the pine forest is well dried by August, and fires spread rapidly once well set. These "dry" thunderstorms sometimes set dozens of fires within a few hours, generally in the highest and least accessible terrain. Summer rain increases to the east and to the south, the mountains of San Diego County being slightly wetter in July and August than those of Los Angeles County.

Some localities in the Peninsular Range are also cooler in winter than one would expect from altitude alone. Cuyamaca, at the altitude of 4670 feet, a thousand feet lower than Mount Wilson, is several degrees cooler, and has just as much snow. Palomar Mountain, very nearly the same in altitude as Mount Wilson, is slightly cooler in January and snowier. Thus residents of California's southernmost counties, San Diego and Imperial, are not so removed from the possibility of winter sports as the modest altitude of the Peninsular Range would suggest. The well-watered summit area of the Peninsular Range is surrounded at lower altitudes by a halo of drier Transitional climate. To the northwest the zone of Transitional climate widens noticeably over the extensive lowlands of the Perris-

Hemet-Lakeview plain. This area, standing only 500 feet higher than the city of Riverside, is removed from the direct path of marine air, and so is warmer in summer, colder in winter, and a little drier than the upper Santa Ana Valley. Despite those differences, citrus plantings have been attempted with good results where air drainage and water supply are adequate. Shortage of water has prevented a full agricultural development of this interesting area. March Air Force Base, which has a long record of very complete weather observations, is located here. Their records show that the Transitional climate, in its coastal phase, is still subject to some marine influence, for at March Field visibility is restricted by fog 5 percent of the hours of April, May, and June, and fog is noted the rest of the year occasionally.

The climate of the lower slopes of the Transverse Ranges is best known through several decades of detailed observations of atmospheric, soil, water, and plant conditions in the San Dimas Experimental Forest, located mainly between the 2000- and 4000-foot levels of the San Gabriel Mountains, above the town of Glendora. At this facility, funded to develop measures to reduce hazards of flood runoff to the valleys below mountain watersheds, hundreds of storms have been charted in detail, with the wind, humidity, and temperature conditions of the long periods of fair weather that intervene in summer. Here the layered condition of the atmosphere over the coast of southern California in summer is revealed by increase upward of summer temperatures, the San Gabriel Divide climatic station at an altitude of 4350 feet being the warmest—the July mean of a 10-year record shows it as warm as 79.4°—whereas San Dimas Canyon station, at 1500 feet, is nearly 7° cooler.

The well-instrumented mountain slopes of the Forest have recorded precipitation much more intense than is shown on the lowlands. At Tanbark Flat (alt.

2725 feet) in a log of 460 storms that took place between 1933 and 1958, it shows that rain has fallen in amounts over 10 inches per 24-hour period on two occasions: a 62-year period of record at the Civic Center at Los Angeles shows maximum 24-hour totals only a little more than half as great. The mean annual rainfall at Tanbark Flat is 28.3 inches, more than enough to fully charge the subsurface capacity of the watershed in an average year; and an intense rainstorm occurring after watersheds have been saturated by antecedent precipitation presents a flood hazard inherent in southern California's terrain and climate. Because of such hazard, the U. S. Forest Service concentrates mainly on watershed protection in southern California. As the watersheds are mostly covered by chaparral, rather than forest, newcomers to the region often wonder why so much concern is evident over mere "brush." However, the fire-flood sequences that have been studied in the San Dimas Experimental Forest leave little doubt about the protective value of the chaparral in this land of long droughts and occasional short violent rains.

Beaumont represents very different circumstances from the highland stations just discussed. Located in a mountain pass site, San Gorgonio Pass, it is reached by gradual ascent, some 100 miles by road from Los Angeles. Its altitude of 2589 feet is sufficient to reduce both summer and winter temperatures below those in adjacent but lower parts of the Intermediate Valley section. Orchards of cherry and apple trees are found in its environs, rather than the subtropical fruits found in lower and warmer sites.

San Gorgonio Pass is one of the major air valves in the circulation of the inland part of the Lowland of Southern California. The cant of windrows, as well as instrumental readings, show that here wind prevails from a westerly direction in most months of the year. In the days of bulky, low-powered, lighter-than-air craft, the journey through San Gorgonio Pass was a matter of concern, for on some days passage could barely be made against the strong west wind. Despite pronounced ventilation of the pass, low clouds and fog from the coast reach inland as far as Beaumont on about 50 days a year, mainly in the summer months. Seldom do they penetrate more than a few miles farther east, however, for in the pass the drier air of the interior mixes with the coastal air, and clears it.

High Desert

Beyond the crest of the Peninsular and Transverse Ranges winter precipitation drops off within 10 or 20 miles to one-third or so of the amounts received on their coastal approaches. Small amounts of summer rain do occur in the interior regions (and do not usually extend to the coast) but are by no means sufficient to make good the deficit of winter rainfall. Where the annual precipitation drops below 10 inches, the climate has been assigned to either the High Desert or Low Desert category, depending upon temperature conditions.

In the High Desert the mean temperature of January is below 50°, July below 90°—much below close to the Transverse Ranges, where the desert floor rises above 3000 feet in altitude and July means are below 80°. Despite its relative coolness in summer, the High Desert seems oppressively hot on a summer day to the traveler from the coast, and winter days will seem chilly. Spring and fall are pleasant, although the wind is often strong here, as it is in most deserts. The lack of vegetation allows active air movement much closer to the ground than is typical in humid climates where the plant cover is profuse.

In terms of human comfort, even the summer afternoons of the High Desert are tolerable to a person in the shade, if water is available. Training exercises with Marine recruits have indicated that summer conditions at Twentynine Palms are no more likely to induce heat prostration than in coastal North Carolina, despite the relative coolness of the latter. The advantage of the desert locality comes, of course, from its superiority in body cooling in the dry air by sweating. However, this creates a tendency toward dehydration that can be countered only by drinking water freely. The need for water cannot be denied through discipline or even inactivity. The summer traveler in desert regions is wise to stay within reach of water, and not to attempt prolonged walks in the sun.

Lancaster and Twentynine Palms represent two versions of the High Desert. Of the two, Lancaster is somewhat cooler, summer and winter. Very low relative humidities prevail in daytime hours of the summer, and seldom at any time of the year does the air feel damp. The virtual absence of a cloud cover or obstructions to visibility allows the air of the High Desert to approach optical perfection as a medium for the flight of aircraft. The famous test center of the U. S. Air Force at Muroc, near Lancaster, has a record of "contact" flying weather (ceilings above 950 feet, visibility more than 3 miles) on 99 percent or more of the hours of all months, on the basis of a 6-year average.

Snow is by no means unknown in the High Desert, and sometimes lightly covers the ground for a few hours or even days. Altitude is critical, and so roads above 4000 feet are sometimes clogged with snow. Under fair-weather conditions, the basins and valleys of the desert receive the cool air that drains from surrounding, higher slopes, as described earlier with respect to the Intermediate Valley climate. Hence, frosty winter mornings may find a skin of ice over standing water on low-lying ground when temperatures have remained above the freezing point on a nearby hill. The coolness of Lancaster compared to Twentynine Palms is explained in part by Lancaster's location on the floor of a large, flat-floored basin. Twentynine Palms is located at the margin of the valley in which it stands.

To the south of Twentynine Palms are the Little San Bernardino Mountains, where the Joshua Tree National Monument has been established. This desert range apparently fails to receive as much as 10 inches of precipitation even on its summits. However, elevations are sufficient to lower temperatures to levels quite comparable to those of Sandberg. Its attraction as a public park lies in the pristine expanses of desert plant life in a mountain setting of moderate relief, with prominent granite boulders. The combination of moistness and heat found in the Little San Bernardino Mountains is unique in southern California. Nowhere else are stands of the Joshua Tree quite as lush or extensive, although they also grow along the foot of the desert slopes of the Transverse Range for more than 150 miles, and elsewhere.

The High Desert merges imperceptibly with the Low Desert as the terrain of the Mojave gradually and irregularly descends toward the Colorado River. Interpolation based on the few places with climatic data permits the assumption that in the vicinity of the 1500-foot contour the July mean increases to more than 90°, and that the January mean rises above 50°. Despite the indefiniteness of the transition zone, it is relatively narrow compared to the areas standing well above it, so that a separation of the Low from the High Desert can be justified.

Low Desert

The Low Desert climatic type is the driest, warmest, and sunniest of California's climates. Most of its stations record less than 5 inches of precipitation annually, all have mean maxima in July that exceed 100°, and the sun shines more than 90 percent of the hours that it is above the horizon.

As in the High Desert, a little rain is received from scattered thunderstorms, but only in the Low Desert is winter rain so deficient that the rain of the summer half-year amounts to as much as 30 to 40 percent of the year's total. Irrigation is an obvious necessity for all cultivated plants, and the fortunate presence of the Colorado River nearby has allowed introduction of water at moderate cost to extensive areas of the Low Desert, chiefly the Palo Verde, Imperial, and Coachella Valleys. Those same valleys share a winter sufficiently warm to allow year-around crop production, the Imperial Valley especially being famous for its winter vegetables.

Many thousands of winter residents and tourists are attracted by the warmth and sunshine of the Low Desert, qualities not appreciated until the success of Palm Springs was clearly demonstrated. With a particularly wind-protected location at the meeting of the valley floor with the San Jacinto Mountains, this remarkable city often enjoys delightful weather when cloud, storm, or wind affect other parts of southern California. A belt of strong wind is often present along the axis of the Salton trough, but various communities have avoided that axis by locating to one side. Interstate 10 cannot avoid the zone of strong winds in the stretch from Garnet to Thousand Palms, where flying sand and gravel often drift across the highway. No other stretch of road in California has given rise to so many claims against insurance companies for the replacement of pitted glass on sand-blasted cars.

In later years other towns, smaller than Palm Springs, have been established in the Coachella Valley on the western shore of the Salton Sea and along the Colorado River, in an attempt to capture a share of the winter tourist trade and even year-around residents. On the latter point, it should be remembered that the sizable recent growth of Arizona's largest cities has taken place in a region having a climate similar to that of California's Low Desert. Great summer heat is not a deterrent to many people in choosing their residence, and is a definite attraction to some. The possibility in this half of the twentieth century of maintaining comfort during the hottest hours by air refrigeration indoors allows full enjoyment of the cooler hours of night and morning outdoors. Furthermore, only the warmer half of the year demands such measures: if only those months with means above 70° require the use of home air conditioning, then Yuma needs it 7 months of the year, no longer than Tampa, Florida, and less than Miami. Furthermore, the dry air of the Low Desert renders supportable higher temperatures than is the case in humid Florida, an advantage not fully illustrated by Yuma in table 2. Chosen for its long and complete climatic record, Yuma's location near the Gulf of California (it was once a river port) brings it air more humid than that representative of the Low Desert area except for the southernmost portion of the Imperial Valley.

The great urban growth of the coastal sector of southern California has led many agricultural enterprises to seek less expensive land. In that search much attention has been paid to the nearby desert as a home for the plant and animal industries whose prosperity enabled Los Angeles County for so many years to place among the leading counties of the nation in agricultural income. Many factors of climate have been considered, frost incidence among them. The High Desert is too cold for most orchard crops, but the frost characteristics of the Low Desert are no more formidable, in many localities, than those parts of the Intermediate climatic region in which certain subtropical fruits have been commercially successful. Lemons, limes, and avocados may never do well in desert environments, but grapefruit is a demonstrated success, and orange production may expand, for example.

Sunshine in abundance, combined with heat, induce growth rates faster than any observed elsewhere in the United States for a responsive plant, such as alfalfa. Should mechanical or chemical capture of solar energy ever become economic for industrial purposes, the climate of the Low Desert will then offer an opportunity unmatched elsewhere in the United States.

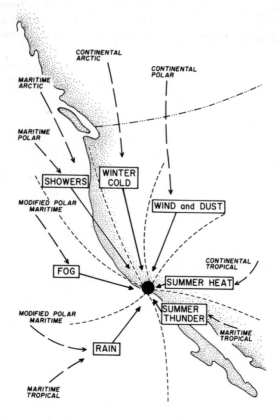

WEATHER TYPES OF LOS ANGELES

CONTINENTAL ARCTIC

CONTINENTAL POLAR

MARITIME ARCTIC

MARITIME POLAR

MODIFIED POLAR MARITIME

SHOWERS

WINTER COLD

WIND and DUST

FOG

CONTINENTAL TROPICAL

SUMMER HEAT

SUMMER THUNDER

MODIFIED POLAR MARITIME

MARITIME TROPICAL

RAIN

MARITIME TROPICAL

Fig. 7. Weather types at Los Angeles related to wind direction and season. Prepared originally by A. K. Showalter.

[46]

WEATHER TYPES

Climate has often been defined as average weather: here this definition is unsatisfactory because it needlessly eliminates the irregular and short-term weather events that are concealed in averages. The cyclical influences of the day and the year would lead us to expect, for example, that it will be cool out-of-doors on a southern California winter evening, but only by actually taking a look outside on a January evening will we determine whether it is clear or cloudy, calm or windy, wet or dry underfoot. Indeed, the fluctuations of daily weather are so important that they often seem to transpose the seasons, making some winter days just like days of summer.

Figure 7 shows certain types of weather in West Los Angeles, with the regions in which air masses have formed and their paths of movement toward southern California. As the wind varies in direction, weather is imported from far-off places, thus bringing to Los Angeles a far greater variety of weather types than would be expected from purely local circumstances. The sensitivity of weather conditions to strength and direction of wind is a matter of greatest importance in weather forecasting.

Wind itself is a variable element of climate, which means it can't serve as a ready-made key to all the details of local climate, but it is helpful, nevertheless, to know in a general way what is to be expected from the circumstances of air flow. Certainly in figure 7 it is clear that fog (also stratus, the low sheet-cloud often called "high fog" in Los Angeles) and rain are associated with circulations from the sea, and that the extremes of winter cold, summer heat (which includes some thundershower activity), and dust are related to movements of air from the land.

Winter Circulations

Rain occurs in southern California most often in connection with large, general storms that move from the sea to the land in the winter months when the circulation of air over the Northern Hemisphere is most vigorous. These storms differ greatly in size, and are represented on the daily weather maps of surface weather by conspicuous centers of low pressure, and on upper air charts by ridges of high pressure, alternating with troughs of low pressure. These manifestations of organized activity generally move from west to east, and initiate predictable weather sequences from the Canadian to the Mexican border.

As the disturbances approach the Pacific Coast, weather is mild and often partly cloudy. With the close approach of the storm the clouds thicken and lower; in a major storm rain will last several days, with only minor interruptions. If a strong surge of cooler air is following, a prominent "cold front" will be shown on the weather map, and the storm typically terminates with an hour or so of very hard rain, perhaps the most intense of the storm. In other instances the rain peters out gradually, but in neither case is the rain followed immediately by cold and clearing weather as is typical of winter storms in the Midwest.

Rather, the usual follow-up of rain in southern California is a gradual shift of the wind from the southwest through the north to northeast, or even east. This change, which takes several days, brings in air that has been over land surfaces for increasingly longer periods before reaching southern California. Generally, the air coming from the northwest is still fresh enough off the sea to bring clouds and moisture to high mountains in its path. This means that whereas the main part of the storm brings most of its rain to the slopes of the mountains facing the coast, showers occur on the desert slopes of the Transverse Ranges in the waning phases of the storm. At such times the clouds are thick over the mountains and stream down over the Los Angeles basin, generally evaporating as they move into the drier air down wind. Thus, the usual windward and lee slope relations briefly reverse, to the benefit of the Joshua Tree, piñon, and juniper woodlands leading from the desert floor to the pine-crested mountain summits.

SANTA ANA WINDS

As noted previously, the most common sequence of events following the passage of a winter storm is for the direction of the wind to proceed clockwise around the compass. A change from northwest wind to north brings continental rather than maritime air to southern California, attended by falling humidity and the disappearance of clouds. Drier air still attends the progression to northeast or east winds, leading to relative humidities less than 5 percent, and even as low as 1 percent. These dry winds from the continent have been called Santa Ana winds, or other similar-sounding names, but refer to the same circumstance: import of continental air marked by low humidity and absence of cloud.

In the typical Santa Ana episode the lowlands of southern California are warmer than any other part of the Southwest, and may become warmer than any part of Texas or Florida. Some periods with Santa Ana winds are chilly, however, even disastrously so for fruit growers; still others bring destructive wind storms.

The possibilities for variations in the strength and character of Santa Ana winds are more easily appreciated if their causes are visualized. One must imagine a vigorous winter storm moving along in the westerlies whose cold front has already crossed the coastline. Air that has pushed behind the cold front over the oceans slows down over the land, and is left stranded a day or two after reaching the Intermountain region as a great mound of air, seen on the weather map as a large region of high pressure. This accumulation of air dissipates by flowing just above the surface into the surrounding areas. As the air has accumulated in the first place over high ground—the average altitude of the Intermountain region being at least a mile—it tends to dissipate by flowing downhill. As it does so, it also subsides, adding to the compressional effects upon temperature. These effects clearly lead to temperature increase as the air travels toward lower ground: compressional warming occurs at the rate of $5\frac{1}{2}°$ per 1000 feet of vertical descent.

The amount of air that accumulates to form the region of high pressure, the location and orientation of that region, and the balance created by motions of the adjacent atmosphere control how fast and how far surface winds will carry, and what their temperature will be at any given point. Most often the dissipation of the "Plateau High" (the term often used to designate the accumulation of air previously referred to) takes place as a broad and irregular stream of air southward between the "banks" formed by the Sierra Nevada on the west, mountains more than two miles high, and the Rockies on the east, equally high. If the flow has nearly dissipated by the time it reaches southern California, its force may be spent over the reaches of the desert, and be little noted on the coast. Perhaps the southward-moving flow of air laps against the Transverse Range in the west, but not the east, in which case the low crest south of Palmdale may be enveloped to bring, a little later, strong winds to the western end of the San Fernando Valley, and later still to the coast at the mouth of Malibu Canyon, Topanga Canyon, or some other relatively low route to the sea. More often the reverse is true: the Santa Ana is stronger to the east. In that case winds are sure to descend Cajon Pass (note the hollow arrows in figure 1B), and to blow across the upper Santa Ana Basin with vigor; still further east the wind finds its way down the long axis of the Salton trough, perhaps down the valley of the Colorado River. Quite often the Santa Ana is first noticed in the west, and then reaches points successively farther east, carrying eventually to Arizona and beyond.

In exceptional circumstances the Santa Ana winds reach gale, or even hurricane, force, the latter only in the mountains. In the first forty years of wind observation in southern California, the most intense windstorm of all took place from November 24 to November 26, 1919, a three-day period following the passage of a storm across the coast in central California. Strong high pressure following the storm was noted, and the accompanying winds set new records at some places, notably on Mount Wilson, where a steady northwest wind held at 90 miles per hour for five minutes, and the average for the hour following was as high as 84 miles per hour. The storm was general over the lowlands too, where northwest winds were measured for five-minute intervals at 60 miles per hour at Santa Monica. As in all windstorms, it can be presumed that gusts peaked at speeds greater than those given by 5-minute averages.

Windstorms in southern California have not been studied as carefully as floods and drought, as they have not caused as much damage, and are less amendable to control. Nevertheless, in still another windstorm of note, January 11-13, 1946, damage to citrus crops alone amounted to 3 million dollars. In this case 60 mile-per-hour winds were again noted in the coastal lowlands, as a result of strong northerly flow that represented the dissipation of a large region of high pressure over the Intermountain region. As would be expected, large differences in temperature existed between lowland

and highland; Burbank at 699 feet elevation had a temperature of 59° at the same time that Tonopah, Nevada (elevation 6093 feet) enjoyed a cool 18°.

Just a year earlier (January 7-9, 1945) a large high pressure system had formed in mid-continent, but that time the air sweeping southward from Canada stayed east of the continental divide. Although it brought freezing temperatures to the shores of the Gulf of Mexico it did not affect southern California. Canadian air masses nearly always sweep southward east of the Rockies, a circumstance which usually prevents freezes in the Southland.

However, once every 10 years or so, frigid air of Canadian origin makes the long, rough journey over the mountainous region which has been termed the Intermountain in this study. An exceptionally large and persistent high pressure system is needed to bring this about. In the freeze of January, 1937, two periods of sub-freezing temperature took place, the first between January 6-11, the second January 19-27. Never since has so much fuel oil been consumed for orchard heating in southern California. Minimum temperatures of 15° were recorded near Redlands, 12° near Imperial; the total number of hours below 27° varied from a maximum of 205 hours to none at all, the localities at each extreme being near Pomona! A total loss of 9 million dollars to growers was attributed to this double freeze. Temperatures do not often descend to levels that low in the average Santa Ana condition

Fig. 8. Selected winter days, 1964. January 3: The cold front passing through southern California brings no rain there, but initiates an extensive period of Santa Ana weather; January 4: Santa Ana weather over southern California, fog over San Joaquin Valley; January 6: Santa Ana conditions persist; nocturnal temperatures below freezing in some districts; January 19: Vigorous cold front brings rain to coastal sector.

SELECTED WINTER DAYS, 1964

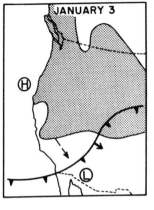

JANUARY 3

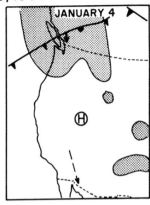

JANUARY 4

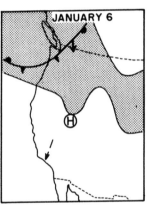

JANUARY 6

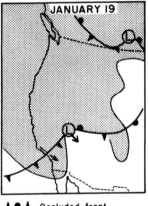

JANUARY 19

▲▲▲ Cold front	●●● Occluded front
●●● Warm front	●▲●▲ Stationary front
→ Direction of frontal movement	⇢ Upper wind direction
Ⓛ Low pressure center	Precipitation in preceding 24 hours
Ⓗ High pressure center	

because the casual high pressure center is usually too small to import air from as far away as Canada. Nevertheless, by the dry air and cloud-free skies the Santa Ana condition contributes to rapid nocturnal cooling, which often causes nightly temperatures to dip at least briefly to sub-freezing levels.

A recent, detailed survey of frost conditions in the Coachella Valley is an example in point. It shows that sub-freezing temperatures occur every year near Indio, a representative agricultural locality with average frost conditions. The last five years studied (1954-1959) showed a total of 50 such occasions. The temperature dropped at least as low as 29° every year, and reached 25° in three of the five seasons. The question arises as a matter of curiosity as to whether or not there is a truly frost-free locality anywhere in southern California. The freeze of January, 1949, ended the frost-free reputation of several districts, among them Niland in the Imperial Valley, and Pacific Palisades, north of Santa Monica (not an agricultural area). The air imported at that time was so cold that the entire atmospheric column over southern California attained sub-freezing temperatures, according to limited radiosonde data. However, the chance for sites of small area to have remained above freezing even at that time, when record lows were being established at most stations in southern California, were greatest on hilly slopes near water. The south-facing slopes of the Santa Monica Mountains, from Hollywood to the sea, may have some truly frost-free sites. The agricultural application of the same principle of selection is seen in the large planting of avocados in the hill country back of San Diego, although they are too far from water to benefit from its presence at times of strong Santa Ana flow. At sea, of course, air temperatures in the Catalina Embayment always remain above freezing, a benefit which presumably extends to the shores of the islands found in it. The temperature at the weather station at Avalon, on Santa Catalina Island, has never fallen below 32° in the 41 years of record, but it is located on a pier. The airport station, 6 miles away, and 1600 feet higher, recorded a minimum of 29° in January, 1949.

Very hot weather is associated with Santa Ana conditions in fall and spring, when the Plateau High is composed of fairly warm air (such a high rarely forms in midsummer, when low pressure is more characteristic of the heated interior). These very warm episodes are even more interesting when the southern California coastal area is the warmest on the weather map. Those used to blaming heat on desert winds are then puzzled, but in well-established Santa Ana conditions temperatures are keyed to altitude, and the High Desert particularly will be cooler than the sweltering coast, the hottest place of all because it is the lowest. Downhill motion, rather than simple movement of heat from some other region, accounts for the warmth of the Santa Ana wind.

The lack of low passes in the Sierra Nevada prevents the Santa Ana condition from being as prevalent in the San Joaquin Valley of California as it is in the southern third of the state. The high pressure conditions in the Intermountain region usually bring calm conditions to the San Joaquin Valley, so calm that air drainage from the mountain perimeter to this great central depression fills it with chill air that turns foggy with the famous "tule fog" of the region. The most pleasant weather of winter in Los Angeles is then often matched by the most miserable in Fresno. This was true in January, 1964, a nearly completely sunny month in southern California, and almost as completely foggy in the San Joaquin.

Although mention has been made of the importance of the summer sea breeze as a spatially variable element of climate, little has been said about sea air in the winter. At that time of the year there is little difference between the temperature of land and sea, and so the thermal basis of the true sea breeze is lacking. Nevertheless, what a weather forecaster calls "gradient winds" account for much air movement from sea to land in winter. This means that the actively moving atmosphere of winter is often arrayed temporarily in patterns that create high pressure at sea, and low pressure over land, irrespective of temperature.

Sea air moving inland does not heat nearly as rapidly in winter as it does in summer, and may even cool if forced to ascend hills or mountains. Also, if sea air invades the land, and the circulation dies down at night, nocturnal cooling may well chill the air to its dew point, resulting in fog. As these fogs form at night over ground, rather than sea, they are called "ground fogs," or "radiation fogs." Such fogs are created in basins and valley floors, localities that have been discussed previously with respect to frost. The same spots that are very cold in dry air are also very foggy in the presence of moist air, when skies are initially clear, the wind quiet, and the nights long. This type of fog is very different from sea fog, which comes about with the chilling of air which originally had such a high dew point that its contact with the sea surface results in widespread fog without the necessity of night-time chilling. The circumstances demand that sea air from a warm environment be imported, and this happens in southern California when a gentle but persistent wind sets in from the southwest or south— gentle, so that turbulence will not cause the fog to lift as a stratus cloud, or high fog; persistent, so that a trajectory can develop with enough reach to bring air of sufficient warmth and moisture. Under these conditions, sea fogs do indeed develop, some of true London type, lasting day and night for perhaps as long as a week with few interruptions. So severe a fog occurs but once in a decade or so, but fogs of shorter duration are much more frequent, as noted in the discussion of the Maritime Fringe. Figure 7 shows sea fogs related to west winds, but the connection is not strong, and the argument is that rain and fog are distinguished more by the manner of air motion than by the direction of motion.

When large masses of air are pulled into the long, upgliding motions that are an internal feature of major winter storms, the stage is set for the precipitation process. It begins with the formation of tiny ice crystals in the upper reaches of the cloud masses that form as a result of the upgliding motions previously mentioned, and is carried through to the formation of snowflakes large enough so that they commence their long, whirling descent into warmer layers of air that transform the snowflakes into drops of rain. Unfortunately, Nature's rain-making process is one which does not often suit the southern Californian environment. Too often the storm structure is absent. Distant storms apparently headed for southern California are frequently blocked by the presence of too high an atmospheric pressure in the region intervening. Thus, when the Santa Ana condition persists with only minor fluctuations for weeks, it serves as a shunting mechanism that prevents storms of northerly origin from taking southerly tracks. Sometimes a storm that appears promising is too weak to bring about effective rainfall; others coming from the sea hold off and eventually go to pieces without much effect. Sometimes all but the top layers of air are in a favorable condition, but the upper parts of the storm are mixed so heavily with dry air that no rain comes about. The opposite condition also occurs: I remember a golden sunset on the beach, when out to sea a massive storm cloud at great elevation sent down a curtain of snow and rain in a line perhaps 100 miles long. Not a drop reached the surface, for in its long fall all solid and liquid moisture evaporated. Small wonder that Will Rogers once defined rain as a miracle in southern California.

The rain storms that do occur sometimes catch the weather forecaster unawares, because a significant proportion take the form of upper air disturbances that have not been previously reported. Called "cold lows aloft," these storms are of especial interest because they are not much affected by underlying terrain. They frequently grow as they move inland, rather than dissipate, and so are a major source of winter precipitation to Arizona and New Mexico, where they account for winter snowstorms of some magnitude. Others more normal in their structure originate with little warning at low latitudes, somewhere between the mainland and the Hawaiian Islands; these storms of southerly origin are a major if infrequent source of moisture to San Diego County and Lower California.

Summer Circulations

Seasonal changes are gradual in southern California, and in a climate where each month of the year brings into bloom a part of the plant cover, even the blossoming and fading of flowers is no sure indication of the seasonal cycle. The change from winter to spring is more subtle than the transition from summer to fall. In March and April rain storms tend to decrease in frequency and intensity, but even in midwinter the tempo of rain is so irregular that lack of rain itself is hardly a matter of note. However, the rise of temperature of the spring months, particularly inland, works against the development of strong Santa Ana

[58]

conditions, and so the decreasing vigor of atmospheric circulation not only reduces rainfall but its aftermath as well — outflow of continental air. So spring assumes a blander regime in which day after day the atmosphere becomes a little warmer, a little more hazy, and day-to-day changes in weather type a little less noticeable.

Surprisingly, early summer is usually quite cloudy, with many days of overcast weather along the coast. April, and even May, are often the cloudiest months of the year close to the sea. In their most complete development, these cloud periods last several days, or even a week or more, in which the sky is completely obscured by a thick, gray overcast — the California stratus so well known in the weather forecasting profession for its unpredictable behavior. This lasts over into June, and may give rise to drizzle, or even light rain. This is a time of year that college professors leave their classes to attend meetings of scientific societies, and the author has noted in a 20-year period when driving on coastal highways to such occasions that he has frequently had to use windshield wipers at least part of the way.

All through the remainder of summer a certain pulsation is present in the advance and retreat of coastal clouds. Every two weeks or so the stratus cover thickens, and forms at night not only over the ocean and coastal strip, but well inland. Shielding the ground from the sun, perhaps completely so for several days at the coastline and for at least a part of the morning inland as far as the mountains, these cloudy periods are also cool periods. As the clouds recede to their normal summer pattern, which brings stratus to the coastal strip during night and early morning hours, temperatures increase inland where skies are clear, setting in motion the summer sea breeze circulation. It is this situation which is regarded as typical, seen in

SELECTED SUMMER DAYS, 1964

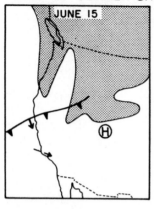

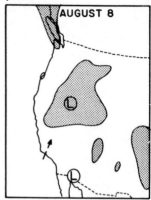

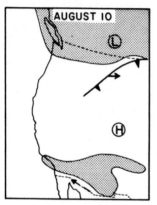

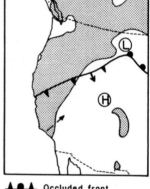

▲▲ Cold front
●●● Warm front
→ Direction of frontal movement
Ⓛ Low pressure center
Ⓗ High pressure center

▲●▲ Occluded front
●▲●▲ Stationary front
— → Upper wind direction
Precipitation in preceding 24 hours

cross-section in figure 4, and which has largely determined the temperature-distance relations shown in figure 5. Similarly, it is the summer situation described for the Maritime Fringe and Intermediate Valley types of climate.

A recent study of the summer climate of southern California relates the alternation of cloudy-cool and clear-warm episodes to distant effects of moving pressure systems farther north in the westerlies, better analyzed at upper levels than at the surface. Its author, Harry Coffin, believes that the changes in cloud cover and temperature that attend the alternating episodes are more likely the result than the cause of such alternation. In principle, he believes that large-scale disturbances in the westerlies affect summer air as well as winter air in southern California, but the summer "storms" bring cloud rather than rain. The details of his argument are not appropriate here, but its larger features are highly significant.

The normal clear-warm situation, which obtains on about three-fifths of summer days (days between May and October), is attended by the dominance of the oceanic anticyclone sketched roughly in figure 3, whereby sea air arrives from the northwest. From this air is drawn the normally thin marine layer which moves inland daily as a sea breeze. Little cool air is transported from the ocean to the land, temperatures rise rapidly from sea to land, and a prominent area of

Fig. 9. Selected summer days, 1964. June 15: Deep flow of marine air across coastline in pronounced Cataline Eddy-type situation preceding weak cold front passage; August 8: Weak sea breeze, great heat inland; August 10: Heat broken by Sonoran circulation, with scattered thundershowers over mountains and deserts; September 2: Early resumption of winter-type circulation, rain in San Joaquin Valley, Catalina Eddy circulation in southern California where intermediate valleys have first cool afternoon since preceding June.

low atmospheric pressure is seen over the very hot lower Colorado River Valley on the daily weather map. With the approach of an oncoming "storm" important changes take place in the distribution of pressure, upper air, and surface winds. These charges are similar in kind, but not in degree to those that accompany a winter disturbance, and because they are relatively subtle we distinguish the word "storm" as it applies to summer conditions in southern California. Farther north along the coasts of British Columbia and Alaska such a summer storm does produce rain, however.

As seen in the Southland, the most prominent feature of the weather change is a tendency for the air crossing the Catalina Embayment to curl around in a counterclockwise swirl, so that sea air crosses the coastline nearly at right angles. This feature forms the cyclonic eddy known for years as the Catalina Eddy, where the marine layer deepens to as much as 6000 feet. The increase in depth of this moist air favors cloud formation within it, which produces the extensive overcast days of late spring and early summer. The cloudy-cool episodes which attend Catalina Eddy conditions average about three days long, but 10 percent last 6 days or more. With them comes not only relief from summer heat inland, but a lessening of fire hazard in the chaparral-covered coastal slopes of the mountains. Temperatures in the desert are less affected, but sometimes drop noticeably when marine air is particularly abundant.

In August and September another type of summer disturbance occurs which approaches from the south, rather than the north. The analysis of the southerly disturbances was first elaborated by Dean Blake, a forecaster on duty at San Diego in the 1930's. He noticed that spells of showery weather approached San Diego in late summer from the southeast, and so named them "Sonoran" storms, after the Mexican state of the same name. Now we know that the Sonoran storms usually have a more distant origin than the Gulf of California, that many originate in events in the Atlantic and Gulf of Mexico.

In figure 3 broken arrows are shown over the continents (except South America) adjacent to the inner margin of the areas of Mediterranean climate. These refer to summertime anticyclonic circulation in the upper air, at heights of from 10,000 to 15,000 feet. Such anticyclones are related partly to circulation over the continents and oceans far to the east. In the case of North America, the great oceanic anticyclone over the Atlantic reaches far west, in the upper air, and commonly overlies the southern part of the North American continent. When farther north and west than usual, this anticyclone is in a position to reinforce a weaker and drier high-level circulation of the same type normally present over the continent. As a result of the reinforcement a tongue of moist air flows aloft toward the southwestern United States, an event visible to a ground observer by the approach of numerous thunderheads from the southeast. Invasions of moist air from the tropical Atlantic and Gulf of Mexico account for most thunderheads forming over the mountains of southern California. The more extensive bring at least sprinkles to the coast. As a matter of fact, I have never seen a summer in which rain did not occur at least once in the coastal sector, and my period of observation reaches over more than 30 years.

It is necessary to add that summer rains are sometimes related to tropical disturbances in the Pacific. In the last century these storms, called *chubascos*, frequently crossed the coastline, but in the twentieth century tropical storms have stayed well out at sea, a difference in behavior which is not understood. Even when many miles offshore, however, chubascos send northward in advance of their presence a deep stream of warm, sultry, and showery air that seems more akin to East Coast than to West Coast summer weather. Only exceptionally do they follow a path close enough to shore to raise heavy winds and seas in the Catalina channel and other coastal points.

September is sometimes a month to be more forgotten than remembered in southern California, for it can turn into a period of exceptionally hot weather marked by stagnant circulation that fails to disperse air pollutants. The combination of heat, smog, and lack of wind is not a happy one, and fortunately one that does not occur often in extreme form. Now and then September turns out to be the hottest month of the year in Los Angeles, but more often the heat is intermittent, and is related to the importation of the Continental Tropical air mass shown in figure 7.

Continental Tropical air in this case refers to the air mass developed in summer in northern Mexico and adjacent parts of the United States, particularly southern Arizona and the lower Colorado River Valley. This very warm region does not often send air toward the coast, but in September it is possible for extremely large, slow-moving anticyclones to develop in the Midwest. When this happens an east wind sets in which brings to southern California a taste of Arizona weather. Unlike the true Santa Ana condition, subsidence within the air mass and a sinking trajectory over the ground are not large components of motion, and so compressional warming adds little to its heat content. The air is hot mainly because it is imported from a hot source region.

Sometime in the first two weeks of October the first intrusion of winter weather occurs. A vigorous storm crossing the coast well north of southern California washes a deep wave of marine air over the coastline, bringing cool temperatures for a day of two, marked usually by clouds indicating active air motion aloft, and a few showers here and there. After this comes the first Santa Ana of the cool season, which in October arrives as a heat wave accompanied by air so dry and so windy that its discomfort is small. Nevertheless, caged animals such as poultry are adversely affected by this, as well as other types of heat. I lost two household pets (white rats caged outdoors) in successive years, within two days of the same calendar date, on occasions of early October Santa Ana winds.

This summary of summer weather includes no mention of destructive wind storms, as they are so very rare. Neither tornadoes nor tropical hurricanes are significant in the recent weather history of southern California. Heat waves are no more extreme than in most other parts of the Nation, and hot humid conditions are felt on only a few days a year.

FIRE, FLOOD, DROUGHT, AND SMOG

Fire, flood, drought, and smog—all topics included are matters of public interest in southern California, and have required legislative and administrative machinery for their treatment; all, furthermore, are intimately related to the climatic characteristics of the region. In a very real financial sense, all are problems in applied climatology.

Fire

Nearly all destructive fires are man-caused in southern California, the only exception being those caused by lightning, which sets fires in forest and brush lands which are well away from heavily populated areas more often than not. Once a fire is out of control, however, the conditions under which it burns are important in the effort to gain control. We have interest, therefore, in the environment of fires, as well as their cause. The concept is held so firmly by the U. S. Weather Bureau that it maintains Fire-Weather Offices in California, from which specialists are detailed to sites of major fires so that fire-fighting crews will have the best possible chance to anticipate changes in weather.

Two kinds of weather, above all, put fire-weather forecasters on the alert. The first is a period in which electrical activity gives promise of lightning-set fires; the second is any Santa Ana episode in which winds are strong, especially in the fall when fuel materials have dried out during the long, warm, dry California summer.

Lightning is a cause of fire chiefly in hill and mountain lands. As discussed previously, thunderstorms tend to increase in frequency with altitude and distance inland. The Angeles National Forest, which is located midway between the extremities of the Transverse Range, is neither the least nor the most affected by lightning of southern California's National Forests. Its lower and most coastal parts average but a few thunderstorms each year, but higher areas farther east record as many as a dozen in the average year. In exceptionally persistent periods with moist air aloft, thunder may be heard each day for as long as two weeks. As 40 percent of days with thunder have a record of fire in the Angeles National Forest, lightning can be counted upon as an efficient cause of brush and forest fires. It has been estimated for California as a whole that nearly 600 fires are set by lightning in mountains in the average year, burning over 56,000 acres.

Unfortunately, rain is so light and scattered from a summer thunderstorm in the Southwest that it does not reliably suppress the fire that the same storm might have started. If it is to be put out, man has to do it. Not to do so runs the risk that the fire will burn on until extinguished by the first general rains of fall. In the coastal lowlands summer thundershowers are even less frequent than those of winter: winter storms connected with the passage of active cold fronts do indeed create lightning, but are usually accompanied by copious rain, and in any event are quite rare (table 2).

The coastal lowlands are by no means immune to Santa Ana winds, however, and it is in the forecasting of their characteristics that most responsibility is thrust upon the personnel of the Fire Weather Office. Any fire is difficult to control that gets a head start in the presence of dry air in gusty, rapid motion. The fire that swept through the hilly Bel-Air district of West

Los Angeles in November, 1961, is an example (see plate 8). Despite the action of 124 city engine companies, 23 tanker units, and the release of fire-retardent solutions from 16 aircraft, damage in excess of $25,-000,000 took place. The Bel-Air fire ranks as the fifth costliest fire in the history of the United States, and the worst in California since the fire in San Francisco that followed the 1906 earthquake.

The circumstances that preceded the fire included the driest rainfall season on record at Los Angeles (4.85 inches), and less than a tenth of an inch of antecedent precipitation in the rainfall season beginning July 1, 1961. A dry, northerly wind set in three days before the fire, bring clear blue skies and high temperatures; on the morning of the fire, November 6, the relative humidity was only 6 percent on the crest of the Santa Monica Mountains, and dropped to 3 percent during the day. At about 8 A. M. the fire started near the crest, nearly directly north of the UCLA campus in West Los Angeles.

The fire spread rapidly in the 50-mile-per-hour winds that raced southward through the canyons toward the city below. Leaves and the woody material of the chaparral were so dry that witnesses described the fire as burning silently, without the usual crackling of burning wood; the fire spread rapidly by wind-tossed burning brands and hours elapsed before effective fire lines could be established. In the meantime, some 484 homes within a 20-mile perimeter had burned, perhaps 20 percent of the expensive hillside and canyon-bottom homes in this part of the Bel-Air district.

Similar destruction has never taken place in the residential areas of Los Angeles on lowlands where homes are widely spaced and separated by lawns and other green vegetation. Wind, humidity, slope, and, above all, the enormous amount of fuel contained in well-dried chaparral create a high fire hazard every year in the brush-covered slopes of southern California; the Bel-Air fire is an example of the realization of that hazard.

FLOOD

From Mission records and other documentary sources that pre-date the instrumental data of the U. S. Weather Bureau, one gathers that floods have occurred in southern California about six times a century. The cause of a flood in this environment is quite simple, free as it is of the complicated factors of frozen ground and heavy snowpacks of colder parts of the country: a flood happens in southern California whenever it rains too much for stream banks to retain the water flowing through them. However, natural water courses in southern California do not have the capacious channels that are typical of drainage basins in humid climates. Although well-incised canyons are to be found in mountains, the streams that issue from them commonly lack well-defined beds on the valley floors upon which they empty. Under natural conditions temporary and wandering streams ran freely over the lowlands after periods of heavy rain. This state of affairs is entirely consistent with geological evidence, which indicates that the major lowlands of southern California have had a structural origin, and have been deeply filled with rock materials swept down from the mountains. In valleys that are basins of accumulation rather than hollows excavated by erosion there is no basis for the formation of deep, permanent, stream-formed channels and if they are to exist they must be created by man, instead.

Still another factor in the behavior of streams is the state of vegetation covering the slopes that drain into valleys. Where the vegetation has been undisturbed for many years, it becomes dense and deeply rooted, with the result that neither floods nor destructive erosion follow rainstorms of ordinary intensity. The fire-flood sequences studies in the San Dimas Experimental Forest, where chaparral is the native vegetation, show marked increase in the peak discharges of streams following rain when only small portions of a watershed have suffered a fire. If as much as a third of the chaparral cover is destroyed, peak flows in one watershed were observed to have increased 67.7 times the normal flow of rain on dry soil, 15.6 times the normal flow on wet soil. The increase in peak flow were also attended by marked erosion in denuded areas, consequently increasing the amount of sediment carried by the swollen streams, and ending with the debris deposited on the more level valley floor below.

About 2 inches of rain fell in the month of November, 1961, on the slopes denuded by the Bel-Air fire, with the same sequence of events observed so often in the San Dimas Experimental Forest leading to substantial flows of mud. However, a much larger demonstration of the same problem had been given by a fire in 1933 that largely destroyed the vegetation cover of Pickens Canyon and adjacent canyons above the communities of Montrose and La Crescenta, a few miles north of Glendale. The watersheds concerned have very high-gradient streams, upon which heavy rain fell during the last three days of December, 1933. After a 14-hour period of continuous rain on December 30, the storm was ended by the passage of an upper air disturbance (an occluded front) near midnight, which created an intense shower lasting about a quarter of an hour. This shower that triggered off not only flood waters, but large movements of debris that had gathered on stream floors for perhaps 30 years (the last major flood had occurred in 1914). Major damage was confined to Montrose and La Cresenta but some streams nearby in areas unaffected by fire also flooded, indicating in this case that the burned area had the immense bad fortune to have been the target of a restricted zone of exceptionally heavy precipitation as well.

Fortunately, not every rainstorm is attended by flood in southern California, for most rain soaks into the ground rather than running over the surface. Even major storms may pass without serious flood, owing to the capacity of rocks underground to store water, which is released gradually over a period of days and weeks. But after abundant rain has occurred in early winter, subsequent storms bring precipitation to watersheds well soaked. Even with a full vegetation cover, as the San Dimas data indicates, peak flows increase as the watersheds approach saturation. This is another way of saying that if it rains enough it's going to flood regardless of other circumstances.

A case in point is given by the flood of 1938, when very heavy rain fell generally over southern California from February 27 until March 4. Although the storm was not record-breaking at many lowland stations, it was in the mountains, where as much as 30 inches accumulated during the storm. This was almost twice as great as the maxima observed in the storm of December, 1933. The presence of the mountains athwart air flow from the south considerably augmented the rain-making process. Stream runoff was so great that it caused the loss of 87 lives, and damage of $80,000,000. In subsequent years flood control measures, such as debris basins, large dams for the retention and regulation of flood waters, and lined channels have prevented a repetition of the disasters of the 1930's.

It is also true, however, that flood-making rains have decreased in frequency and severity in recent years. In the 20 years between 1944 and 1963 precipitation has only once exceeded 20 inches at Los Angeles (1952 with 24.95 inches), whereas in the first half of the precipitation record (1878-1920) totals at Los Angeles exceeded 20 inches in 11 out of 43 years, approximately 25 percent of the time. We cannot predict that rainfall in southern California will not again resume its former pattern, and so concern over flood remains, despite the unpredictability of the time and place in which its effect will be felt. The enormous increase in population that has taken place since the 1930's increases the hazard, in the sense that vulnerability to flood is greater.

DROUGHT

Not only have there been fewer very wet years in Los Angeles recently, but the number of very dry years has increased. If the 86-year record of precipitation measurement at the city station is divided into two halves, and if annual precipitation less than 10 inches is defined as a drought year, then 10 drought years occurred between 1878 and 1920, and 13 from 1921 to 1963. The last 20 years of record include 7 such dry years; between 1944 and 1963 fully three-fourths of the years have brought less than 15 inches of rain annually. Inevitably, the decline of wet years and the increase of dry years have lowered the long-term mean. The first 43 years of record gave a mean annual precipitation of 15.52 inches, the second 43 years 14.76 inches, and the last 24 of those 43 years only 11.94 inches.

There is no reason to think that the Los Angeles area is not typical of southern California as a whole. Its record establishes the basis for the conclusions that (1) rainfall has gradually if irregularly declined in southern California since instrumental records have been kept, and (2) a drought of major magnitude has taken place in the last 20 years.

Several things should be said concerning the points

just made. In the first place, the underlying physical reasons for fluctuations in rainfall are not understood any better than are the widely noted recessions of glaciers in the twentieth century, or the warming trends in both the North Atlantic and North Pacific. If neither the type of change taking place nor the rate at which it has taken place has exceeded natural limits, there should be droughts on record at Los Angeles equivalent to the present. That of the 1890's is well known; although shorter than the current drought, its driest 10 years (1894-1903) averaged but 11.1 inches, equal to the 11.1-inch mean representing the driest 10 years of the current drought (1953-1962). Earlier droughts are known less exactly, but from old documents, it seems clear that rain in the years between 1785 and 1810 was generally below normal, as also between 1819 and 1833.

In the next place, droughts do not seem to correspond well enough to sunspots, or other physical phenomena that are cyclical in occurrence, to allow us to forecast them ahead of time. Similarly we do not know when the present drought will end, or even if it will end.

Finally, it is necessary to distinguish between a state of water shortage and one of drought. A society is short of water when it does not produce enough to satisfy its demands, and this shortage can occur in the midst of plentiful rainfall if the supply to consumers is insufficient. Drought, in contrast, has nothing to do with demand, and everything with supply: it refers to to a condition in which the atmosphere supplies less rainfall than normal. It is unfortunate that the growth of southern California in the postwar era has taken place at a time when only about 75 percent of normal rainfall has been received. However, it should be realized that even in the past demand outstripped local supplies of water, so variations in availability of local water no longer make the whole difference between having enough and needing more. The city of Los Angeles began importation of water in 1913; by 1947 a survey of the ground water conditions showed overdrafts to be widespread in southern California: rural as well as urban areas were using more water than was being replenished by natural inflow. The current drought, then, emphasizes the water shortage of southern California, but its relaxation, when and if it occurs, will only lessen, not cure, that shortage. More water is still needed.

The only source of water increase that is meteorological in nature is the increase in rainfall that sometimes attends the seeding of clouds with dry ice, silver iodide particles, water droplets, or other substances introduced into clouds from aircraft or carried by air currents from the ground. Careful experimentation in California has not shown conclusively that rainfall is increased by such artificial methods, although in certain storms little doubt exists that stimulation of precipitation has been achieved. In other storms, however, rainfall seems to have been actually inhibited. Experience and experimentation, though, will help us learn which terrains and storm situations are best suited to cloud seeding. In any event, the great limitation is that "rain-making" can only be done when it is either raining naturally, or is about to rain. Therefore, no weather modification yet proposed can hope to produce rain in dry periods when it is most needed.

At present it seems that only continued importation of water will meet the needs of southern California, with even larger importations in the future. The market for partially or completely de-salinized water from the sea will grow as costs of water from other sources becomes more expensive. Over-all, though, water will remain so cheap in comparison to other costs of living that its relative unavailability in southern California should not inhibit increase of its urban population.

Smog is a word that has had its origin in the United States in the twentieth century, but might have come out of England much earlier to describe the mixture of fog and coal smoke that has long hung over London and other large British cities. Man creates air pollution, but the atmosphere makes it less evident depending on how efficiently it is dispersed from its source of origin. Thus, the smog potential is regarded as low where smog is ordinarily dispersed rapidly, but is said to be high where dispersion rates are low.

Coal smogs are thickest in basins or valley floors on calm nights of winter, especially if there is a snow cover. At such times more coal is being burned, and cooling is intensified by air draining from higher surroundings. Before the use of bituminous coal was curbed, now and then downtown Pittsburgh was overrun in such a manner. Cold bottom air draining down the slopes of the Allegheny Plateau picked up smoke from thousands of household chimneys, as well as from the factory smokestacks along the highly industrialized river banks upstream from the city. Before the 1940's no Angeleno thought of his city on the same terms as the "Smoky City," and to him the term smog was unknown or unfamiliar.

During World War II, however, the air in the vicinity of Los Angeles underwent so great a change in quality that the city that had started the decade unconscious of smog required by 1948 an Air Pollution Control District for the analysis and supervision of sources of pollution. But carbon soot and sulfur dioxide, the products of coal combustion, have never been the principal components of Los Angeles smog, and, obviously, a calm spell in winter snow is not its setting. A new smog, it is apparent, was born in the southern California environment.

The new smog has been termed an oxidation smog.

It is the product of petroleum combustion in the presence of sunlight and relatively still air, which has occurred in southern California as nowhere else. However, other large areas of the southwestern United States which are also sunny and overlain by still air many hours of the year run the danger of the same oxidation smog should sufficient combustion take place.

The effluent of the combustion engine—a mixture of nitrogen dioxides and hydrocarbons—is acted upon by sunlight to form ozone, which despite its popular reputation as a healthful substance causes eye irritation, plant damage, and reduction of visibility. The complete profile of the substances present in smog is a complex subject which is still undergoing investigation, but oxidant content is at present most widely used as a measure of smog intensity. The parts of ozone per million parts of air frequently rise to as much as one-half part per million in smoggy air, whereas normal surface air contains about one-twentieth part per million. The ten-fold increase of ozone affects visibility by the creation of haze-forming particles larger than those in normal air; the presence of other smog-related substances often gives the haze a wellowish or amber tinge quite unlike milky sea haze.

Diffusion upward is often inhibited in southern California by the kind of temperature inversion illustrated in figure 4, particularly in the warmer months, the time of the year in which air is shifted mainly as a light sea breeze. The sea breeze moves at a normal speed of from 8 to 16 miles per hour, but it is preceded by a quiet period of several hours in which it is charged from sources of pollution without the benefit of dilution through large-scale motion. In September the quiet period preceding the sea breeze includes the morning rush hour, and the beginning of the daily work period; only then, perhaps at 9 or 10 A. M., does a substantial movement of air take place across the coastline. On a bad day, when the base of the inversion is only a few hundred feet above the ground, heavy smog will be forecast. Drivers are then urged to reduce their travel as much as possible, and industry instructed to switch to natural gas as a fuel in place of petroleum. Wind studies show that with few exceptions air passes through the coastal sector within a single day, which is to say that the terrain between the coastline and mountains receives a change of air every 24 hours. Thus the smog cycle is a daily one, in essence, and does not involve the addition of pollutants to air contaminated during the previous 24-hour period. It shows a definite peak in daytime; the worst hours commonly occur between noon and 5 P. M.

The area most affected by smog is the Los Angeles area itself, and communities surrounding it, particularly those lying to the north and northeast within a range of 20 miles or so. If it seems surprising that contamination lasts so far downwind from the source area, the qualification should be added that sources of pollution are widely scattered, and are concentrated in south and central Los Angeles only in a relative sense. Also, smog is manufactured as the contaminants are traveling with the wind by the photochemical processes described before, and so its dispersion is not at all comparable to diffusion of soot particles from a smokestack.

As viewed from aircraft, on smoggy days, the entire coastal sector of southern California is seen to be covered by a thick haze. The estimate has been made that 97 percent of the population of California lives in areas in which visibility is impaired by smog, and the great majority live close enough to sources of pollution potent enough to cause at least occasional eye smarting.

Individual sensitivity to smog varies greatly, and most people become accustomed to prevailing oxidant levels without awareness of eye irritation or other effects. Heavy smog is always noticeable, however—all the more so because there are many weeks, sometimes months, when smog is of little consequence, and often completely absent. It is possible that the worst days of smog, as they were experienced in the 1950's, will not be repeated owing to advances in control of industrial wastes, and even more definite improvement is expected when effective devices are installed to control the emission of fumes from automotive crankcases and exhausts.

Advances in smog control have been handicapped to a degree by the general unwillingness of the California resident to admit that his environment has been invaded by anything that detracts from its physical attractiveness. The scattered and irregular occurrence of bad smog days leads to easy dismissal of the whole affair, and yet it seems quite certain that without the vigorous control that has been exercised by air pollution officers smog would be much more of a nuisance than it is.

FURTHER ACTIVITIES

This survey of the climate of southern California has discussed generally the kinds of weather that are characteristic of the several parts of this region in its several parts. If curiosity has been aroused, there is no need for it to die because the end of this book has been reached. Several avenues are open to continuing interest in the weather and climate of California.

For example, many persons find value in treating daily weather as the building block of climatology. The news media of today—newspaper, radio, and television—make it possible to keep in touch easily with daily weather on the national and local scenes. The character of the weather of months and seasons just passed is summarized in the *Monthly Weather Review*, published by the U. S. Weather Bureau, also in *Weatherwise*, a magazine issued by the American Meteorological Society specifically for lay interest in weather and climate. The *Daily Weather Map*, available by subscription daily from Washington, D. C., will throw light on weather events that are only a few days old (it arrives by mail in California a few days after time of observation).

It is also interesting to many people to carry out some form of weather observation on their own. The activity can run all the way from the role of serious, permanent observation in cooperation with the U. S. Weather Bureau to the operation of a home observatory just for fun, or occasional observations of other types. Many photographers, for instance, develop an interest in the different types of clouds to be seen in the California sky, which tell a great deal about the processes taking place thousands of feet above the ground. A collection can quickly be accumulated that goes beyond the examples that illustrate this little volume.

A reading interest is perhaps the most nearly universal of all, and this can encompass a wide range of materials from the elementary to the technical. A short, annotated bibliography is presented as a suggestion of the variety of sources from which knowledge of the weather and climate of southern California can be extended.

SUGGESTED READING LIST

A. Materials in print. These are available in many libraries, and can be purchased from the publisher through any bookstore.

John Aldrich and Myra Meadows (1962), *Southland Weather Handbook*, Brewster Publications, Los Angeles. 51 pp.
A large paperback, containing useful compilations of statistics, information about weather instruments, forecasts, and brief remarks on weather types of southern California. MB 28. 322 pp.

George Kimball (1960), *Our American Weather*, Midland, A paperback version of a book published in 1955 by the Indiana University Press, Bloomington. It describes the weather month by month for the entire conterminous United States, and in so doing places California in its proper national context.

Clifford Zierer (editor, 1956), *California and the Southwest*, John Wiley & Sons, New York. 376 pp. Chapter 4 on Weather and Climate, written by John Leighly, describes weather types of summer and winter, illustrated by weather maps of considerable detail. Suitable for college-level readers.

Glenn Trewartha (1954), *An Introduction to Climate*, third edition, McGraw-Hill, New York. 402 pp. A textbook giving the elements of climate as well as its world distribution. It has a section on the summer-dry subtropical (Mediterranean) climate.

U. S. Weather Bureau, various publications.
All publications currently in print are listed in Price List 48, available upon request from the Superintendent of Documents, Washington, D. C., 20402, issued annually. References to climatic data are included.

B. Materials available in major or research libraries.

Leo Sergius *et al.* (1962), *The Santa Ana Winds of Southern California*, Weatherwise, Vol. 15, No. 3, p. 102.
A short but well-written account of Santa Ana winds; an account of the Bel-Air fire is included.

James Taylor (1962), *Normalized Air Trajectories and Associated Pollution Levels in the Los Angeles Basin*, Air Quality Report No. 45, Los Angeles County Air Pollution Control District. 49 pp. The trajectories of air parcels across the Los Angeles Basin are determined statistically, not subjectively. The best information yet presented on this difficult subject.

Robert L. Weaver (1962), *Meteorology of Hydrologically Critical Storms in California*, U. S. Weather Bureau, Hydrometeorological Report, No. 37, 207 pp. This report reviews the major storms on record, including local, intense storms not previously analyzed.

Harry Coffin (1961), *Marine Influence on the Climate of Southern California in Summer*, Ph. D. dissertation, Dept. of Geography, University of California, Berkeley. Available only in microfilm, this study examines the causes of cloud and rain in California summer weather.

Bert Bolin (1959), *The Atmosphere and the Sea in Motion*, The Rockefeller Institute Press, New York. 509 pp. Morris Neiburger, in pp. 158-169, wrote on "Meteorological Aspects of Oxidation Type Air Pollution," where he presents the results of several years research in the mid-1950's on smog formation and dispersal in the Los Angeles basin.

James Edinger (1959), *Changes in the Depth of the Marine Layer Over the Los Angeles Basin*, Journal of Meteorology, Vol. 16, No. 3, pp. 219-226. An analysis of 116 flights made by a well-equipped airplane over the Los Angeles basin in the summer of 1957.

U. S. Forest Service, Data from the San Dimas Experimental Forest on (1959) *Mountain Temperatures*, Misc. Paper No. 36. 33 pp.; (1959), *Four Hundred Sixty Storms*, Misc. Paper No. 37. 101 pp.; (1954) *Fire Flood Sequences*, Technical

Paper No. 6. 29 pp. These studies are excellent examples of the analysis of field data representative of hill and mountain environments.

United States Weather Bureau (1959), *Climates of the States— California,* Climatography of the United States No. 60-4. 37 pp. The most convenient assemblage of data available for the state as a whole; brief remarks about terrain, air flow, and storms precede the data.

State Water Resources Board (California, 1955), *Weather Modification Operations in California,* Bulletin No. 16. 271 pp. Analysis the effect of seeding on about 200 storms, 1950-1952, in 3 areas.

Roderick Peattie, (editor, 1946), *The Pacific Coast Ranges,* the Vanguard Press, New York. 402 pp. The section on climatology, written by R. J. Russell (pp. 357-379), includes a skillful description of the major features of the climate of southern California.

Harold C. Troxell (1942), *Floods of March 1938 in Southern California,* U. S. Geological Survey, Water-Supply Paper 844, 399 pp. A thorough analysis of the most destructive flood in the history of southern California.

Richard Joel Russell (1953), *Climates of California,* University of California Press, Berkeley. 11 pages, 1 colored map. This classic work is an excerpt from the University of California Publications in Geography, Volume 2, No. 4. It presents a map of climatic types modified from the Koppen system of classification, and a short discussion of it.

Monthly Weather Review (1934), Vol. 62, March: Lawrence Daingerfield, *The Excessive Rain and Flood in the Los Angeles, Calif., Area,* pp. 91-94. George French, *Meteorological Conditions Attending the Heavy Rainfall in the Los Angeles, Calif. Area December 30, 1933 to January 1, 1934, inclusive.*

Monthly Weather Review (1919), Vol. 47, January: Ford A. Carpenter, *Southern California Windstorm of November 24-26, 1918,* p. 26-27. W. P. Hoge, *Terrific Windstorm of November 24-26, 1918, on Mount Wilson, Calif.,* p. 28. An account of the most severe windstorm in the first 40 years of instrumental records in the area.

Alexander McAdie (1903), *Climatology of California,* U. S. Weather Bureau, Bulletin L. The only monographic treatment of the climate of the state as a whole.

TABLE 1. Definitions of the Climatic Types of Southern California

Mean annual range of temperature	Mean annual precipitation	Mean daily temperature of the coldest month of the year (C) > 50°	Mean daily temperature of the coldest month of the year (C) < 50°	Mean daily temp. of the warmest month of the year
Coastal Climates				
< 30°	< > 20"	MARITIME FRINGE		< 72°
	< > 20"	INTERMEDIATE VALLEY		> 72°
< > 30°	10" to 20"		TRANSITION A	Summer variable depending upon altitude
	> 20"		MOUNTAIN	
Interior Climates				
> 30°	10" to 20"		TRANSITION B	< 90°
	< 10"		HIGH DESERT	> 90°
	< 10"	LOW DESERT		

TABLE 2. Climatic Data for Selected Stations

No.	Station	Climatic Type	Lat. N.	Long. W.	Height Feet	No. of Years*
1.	Santa Maria	Maritime Fringe	34°54′	120°27′	238	30
2.	Los Angeles Airport	Maritime Fringe	33°56′	118°23′	99	30
3.	San Diego	Maritime Fringe	32°44′	117°10′	19	30
4.	Burbank	Intermediate Valley	34°12′	118°22′	699	30
5.	Riverside**	Intermediate Valley	33°57′	117°24′	820	16
6.	Sandberg	Transition	34°45′	118°44′	4517	30
7.	Beaumont	Transition	33°56′	116°56′	2589	13
8.	Mount Wilson	Mountain	34°14′	118°04′	5709	12
9.	Lancaster***	High Desert	34°42′	118°08′	2352	7
10.	Twentynine Palms	High Desert	34°08′	116°03′	1990	28
11.	Yuma (Arizona)	Low Desert	32°40′	114°36′	199	30

Mean Temperature—°F

No.	Jan.	Feb.	Mar.	Apr.	May	June	July	Aug.	Sept.	Oct.	Nov.	Dec.	Year
1.	50.4	51.8	54.0	55.8	57.7	60.1	62.2	62.4	62.1	60.0	56.4	52.4	57.1
2.	53.2	54.1	56.5	59.2	62.1	64.7	67.6	68.2	66.7	63.4	59.3	55.6	60.9
3.	54.9	56.3	58.2	60.5	63.2	65.6	69.3	70.3	68.7	65.0	60.8	56.9	62.4
4.	52.6	54.0	56.9	60.5	64.0	67.5	73.2	73.6	71.2	65.1	59.9	54.7	62.8
5.	51.9	53.6	56.4	60.2	64.3	70.1	75.8	75.9	69.6	64.7	58.3	52.9	62.8
6.	39.8	40.8	44.7	50.5	56.7	64.8	73.9	73.5	69.4	58.3	49.5	42.4	55.3
7.	46.5	47.9	50.3	55.8	61.7	67.1	75.5	74.9	71.8	63.0	54.5	48.7	59.8
8.	41.0	41.0	42.2	49.5	55.8	61.8	71.6	70.5	67.9	57.5	49.4	43.3	54.3
9.	41.3	46.7	50.0	60.4	67.0	73.5	81.8	79.2	74.2	62.4	49.6	43.3	60.8
10.	48.1	52.4	57.2	65.6	73.2	81.6	88.3	86.8	80.6	68.7	56.5	50.0	67.4
11.	55.3	60.1	65.7	72.8	80.4	87.8	94.6	93.7	88.3	76.4	64.2	57.1	74.7

Mean Rainfall—Inches

No.	Jan.	Feb.	Mar.	Apr.	May	June	July	Aug.	Sept.	Oct.	Nov.	Dec.	Year
1.	2.8	2.5	2.3	1.2	0.3	0.1	T	T	0.1	0.6	0.9	2.7	13.5
2.	2.0	2.8	1.9	1.0	0.3	0.1	T	T	0.2	0.4	1.1	2.6	12.4
3.	1.7	2.3	1.5	0.8	0.3	T	T	0.1	0.2	0.6	0.8	2.6	10.9
4.	2.4	3.0	2.2	1.2	0.3	0.1	T	T	0.3	0.5	1.0	2.9	13.9
5.	1.8	2.6	2.0	0.9	0.2	T	T	0.2	0.1	0.6	0.9	2.7	12.0
6.	2.3	2.8	1.8	0.7	0.2	T	T	0.1	0.2	0.7	0.7	2.9	12.4
7.	2.5	2.8	2.8	1.8	0.2	0.1	0.2	0.4	0.5	0.8	1.8	3.2	17.1
8.	5.7	6.9	6.6	2.8	0.4	0.1	T	0.2	0.1	1.5	4.1	7.0	35.4
9.	1.0	0.4	1.1	0.3	0.1	0.0	T	T	0.1	0.3	0.5	1.4	5.2
10.	0.6	0.3	0.3	0.1	T	T	0.5	0.7	0.4	0.5	0.3	0.5	4.2
11.	0.3	0.3	0.3	0.1	T	T	0.2	0.5	0.6	0.3	0.2	0.6	3.4

* For mean temperature and rainfall only; other data usually based on shorter periods

** Cloud, humidity, snow, and thunder data from March Air Force Base

*** Cloud, humidity, snow, and thunder data from Edwards Air Force Base

T Precipitation on record, but with a mean less than 0.05 inches.

| No. | Temperature | | | | Relative Humidity | | Mean Annual Days Sunrise to Sunset | | | | | Mean Annual Snowfall (Inches) |
| | Daily Range | | Extremes | | | | | | | | | |
	Jan. °F.	July °F.	Max. °F.	Min. °F.	Jan. %	July %	Clear	Partly Cloudy	Cloudy	Rain	Thunder	
1.	24.7	20.9	104°	22°	73	78	172	114	79	48	2	T
2.	20.3	13.6	108	23	68	75	137	113	115	37	3	T
3.	18.8	12.4	104	29	68	75	146	123	96	44	3	T
4.	25.3	28.3	111	21	62	63	186	103	76	39	4	0.3
5.	28.2	36.9	118	19	57	50	209	67	89	40	5	0.1
6.	12.4	22.4	102	3	59	25	210	80	75	40	5	27.0
7.	21.0	37.6	110	11	51	44	184	95	86	49	9	2.4
8.	14.4	26.8	98	10	46	29	—	—	—	44	—	49.2
9.	29.7	35.2	113	7	62	31	247	72	46	22	5	4.0
10.	26.7	32.6	116	11	38	19	273	64	28	—	—	1.0
11.	23.9	26.2	120	28	41	34	255	65	45	14	7	0.0